The Morals of Life

INSUBORDINATIONS: ITALIAN RADICAL THOUGHT

Lorenzo Chiesa, series editor

1. *The Monopoly of Man*, Anna Kuliscioff
2. *Convention and Materialism: Uniqueness without Aura*, Paolo Virno
3. *On Freud*, Elvio Fachinelli
4. *Italian Operaismo: Genealogy, History, Method*, Gigi Roggero
5. *The Morals of Life: Biology, Biopolitics, Bioethics*, Davide Tarizzo

The Morals of Life

Biology, Biopolitics, Bioethics

Davide Tarizzo

The MIT Press / Cambridge, Massachusetts / London, England

The MIT Press would like to thank the anonymous peer reviewers who provided comments on drafts of this book. The generous work of academic experts is essential for establishing the authority and quality of our publications. We acknowledge with gratitude the contributions of these otherwise uncredited readers.

This book was set in Arnhem Pro by Westchester Publishing Services. Printed and bound in the United States of America.

Library of Congress Cataloging-in-Publication Data

Names: Tarizzo, Davide, author.
Title: The morals of life : biology, biopolitics, bioethics / Davide Tarizzo.
Description: Cambridge, Massachusetts : The MIT Press, [2024] | Series: Insubordinations: Italian radical thought | Includes bibliographical references and index.
Identifiers: LCCN 2023045517 (print) | LCCN 2023045518 (ebook) | ISBN 9780262549035 (paperback) | ISBN 9780262379625 (epub) | ISBN 9780262379632 (pdf)
Subjects: LCSH: Life. | Biopolitics. | Bioethics.
Classification: LCC BD435 .T187 2024 (print) | LCC BD435 (ebook) | DDC 174.2--dc23/eng/20240317
LC record available at https://lccn.loc.gov/2023045517
LC ebook record available at https://lccn.loc.gov/2023045518

10 9 8 7 6 5 4 3 2 1

Contents

Series Foreword / ix
Preface / xi

Introduction: The Morals of Life: On Metaphysics / 1

I The Morals of Selection: On Biology / 21

1 How Darwin Changed Philosophy / 23
1.1 The Discovery of Life / 23
1.2 The Role of Ignorance / 25
1.3 The Chain of Being / 27
1.4 Life as a Quantity / 30
1.5 Disillusionment / 32

2 How Darwinism Changed Science / 35
2.1 Explanation / 35
2.2 Teleonomy / 37
2.3 *Vera Causa* / 40
2.4 Adaptive Qualia / 46
2.5 The Fit of Form and Form / 52
2.6 Negative Selection / 60
2.7 Scientific Mythology / 65

3 Dogmatism, Scientism, and Critical Naturalism / 73
3.1 From Naturalism to Scientism / 73
3.2 Historical Limits / 75

3.3 *Nexus Finalis* / 78
3.4 Ontological Fallacy / 80

II The Morals of Behavior: On Biopolitics / 85

4 Toward a New Biopolitical Regime / 87
4.1 The Unanswered Question / 88
4.2 Economistic Background / 92
4.3 Epidemiological Apparatus / 95
4.4 Ideological Order / 103

5 From Normalization to Optimization / 109
5.1 Normalization / 110
5.2 Optimization / 114
5.3 From Individuals to "Dividuals" / 118
5.4 Operational Definitions / 122
5.5 From Freedom to Depression / 130
5.6 Foucault and Ethopolitics / 135

6 The Government of Modular Living Beings / 139
6.1 Population of Entities / 140
6.2 The Axiomatic Procedure / 141
6.3 Radical Behaviorism / 144
6.4 Behavioral Selectionism / 148
6.5 A Technology of Behavior / 155
6.6 Optimization and Modularity / 163
6.7 Depersonalization and Disembodiment / 170
6.8 Mythic Metonymy / 174
6.9 Exit / 182

Conclusion: The Moral of Morals: On Bioethics / 185

C.1 Fetus: Life without Body / **186**

C.2 Wannabe: Body without Life / **195**

C.3 Pandemic: On Transmodernity / **207**

Notes / **225**

Index of Names / **263**

Series Foreword

Insubordinations are creative and innovative double negations. They occur when an existing negative condition, the state of being "sub" or "under" a given order and thereby having an inferior rank, is countered by negating this very subjection. In our current late-capitalist predicament such a reversal acquires a more complex meaning. The ordering authority is in fact no longer simply in crisis and exposed to resistance but profoundly disordered in its own operative structure. Today, powers traditionally devoted to regulation perpetuate and reinforce their effectiveness by continually deregulating themselves. Orders become more and more oppressive precisely as they unveil the inconsistency on which they rest. As Pier Paolo Pasolini presciently put it almost fifty years ago, by now, "nothing is more anarchic than power." In this desolate scenario, actual insubordination cannot but arise as the tentative search for a *new* kind of order. Its long-term and admittedly ambitious mission is the establishment of a society without subordinates, called "communism." Its first and more realistic task is a taxonomic critique of an Order that resolves itself into myriad conflicting, yet no less tyrannical, suborders.

The present series aims to dissect the contemporary variant of the double negation involved in insubordination through the privileged prism of Italian radical thought. Starting from the late 1970s, Italy emerged as a laboratory for test-piloting the administration of the state of exception we are now living on a planetary level, both geopolitically and in our everyday lives. A brutal repression put an abrupt end to an intense season of social and political emancipations. But the theoretical elaboration of that defeat, which should not be confused with a grieving process, has managed to promote Italian radical thought to the center of a series of international debates that endeavor to define a new function and field of revolutionary politics. The series moves from the assumption that while so-called Italian Theory remains a vague and awkward category and attempts at hegemonizing it run the risk of resurrecting the idea of a national philosophy, it is beyond doubt that a growing number of left-wing Italian authors have, for good reasons, become very popular worldwide.

Drawing on philosophy, political theory, psychoanalysis, architecture, art history, anthropology, sociology, economics, and other fields, this interdisciplinary series intends to both further investigate consolidated Italian theories of emancipation and introduce authors (both present and past) who still remain largely unknown among Anglophone readers. Insubordinations: Radical Italian Thought will also foster original critical readings that pinpoint the tensions inherent to the oeuvre of prominent progressive thinkers and develop novel dialogues with various strands of post–World War II militant thought (such as heterodox Marxism, biopolitical theory, feminism of difference, social psychoanalysis, antipsychiatry, and theories of Fascism). The series will also translate works by seminal earlier Italian authors who may be regarded as "forerunners" or critics *avant la lettre* of current trends in Italian radical thought.

It is my hope that, by delving into the titles of this series, readers will be able to appreciate the disciplined indiscipline they all share.

Lorenzo Chiesa

Preface

Today, Life is among the most widely used and abused concepts. Interestingly, however, the very definition of Life remains subject to controversy among scholars—so much so that some have concluded that there are as many definitions of Life as there are people trying to define it. For biologists, this news is annoying. As one of them has remarked, "How are we going to discuss it if you believe that the definition of life has something to do with DNA, and I think that it has something to do with dynamic systems? We cannot make artificial life because we cannot agree on what life is. We cannot find life on Mars because we cannot agree what life represents."[1]

Here, I will not ask myself "what life represents" or how to find the true definition of Life. I will not do this for several reasons. First, I am not a biologist. Second, as a philosopher, I basically agree with Carol Cleland that the search for such a true definition of Life is useless and even bad for science because it might keep researchers from attaining a deeper understanding of what it means to be alive.[2] Third, I also find myself in agreement with Paul-Antoine Miquel when he stresses that there might be no such thing as Life itself in the natural world.[3] Four, based on all this, the most urgent question from my point of view is not "what life represents," or what Life genuinely is, but rather what people *believe* Life is, and why Life itself has become an issue of the utmost importance for contemporary society.

Having said this, it is equally important to emphasize from the outset that even though the concept of Life remains controversial, the concept has nonetheless some characteristics on which most scholars agree these days. To begin with, both scientists and philosophers usually share the critical assumption that the concept of Life should be distinguished from the concept of living beings. However trivial this assumption may seem, it is not as obvious as we normally think, and it is not without consequences for our everyday life. Another widespread assumption is that Life as distinct from living beings is to be thought of as both a *process* and a *capacity*. The idea that Life is a process is closely intertwined with the idea that it should be distinguished from any particular living thing. As John von Neumann, one of the founders of the field of Artificial Life, famously put it, "Life is a process which can be abstracted away from any particular medium."[4] With regard to the characterization of Life as a capacity, the accepted notion is that living beings can be recognized by their capacity for self-preservation, self-reproduction, and evolution. According to the National Aeronautics and Space Administration (NASA), for example, the search for Life on other planets is nothing other than the search for "a self-sustaining chemical system capable of Darwinian evolution"[5]—namely, matter that has the capacity to reproduce itself and evolve as survival dictates. This widely known definition of extraterrestrial Life is clearly molded into the definition of terrestrial Life given by the vast majority of evolutionary biologists. As succinctly stated by the neo-Darwinian biologist John Maynard Smith years ago, "Life should be defined by the possession of those properties which are needed to ensure evolution by natural selection."[6]

Thus, although nobody today can say with absolute certainty what Life is, or "what life represents," the concept is most often marked by some unmistakable features: Life is a process of and a capacity for self-preservation, self-reproduction, and evolution. Neo-Darwinism has exerted a paramount influence on this view of Life, as I argue next.

If neo-Darwinism has become so relevant to us, this is not only because today's biologists generally take the Darwinian view of Life and evolution for granted.[7] The reason is also that evolutionary theory is the only branch of knowledge that is entitled to answer the key question: What is man? With the weakening of religious beliefs and the decline of traditional metaphysical doctrines in Western countries, this question—which is crucial for determining how to deal with human beings and govern human societies—has been left to empirical science, in particular the science that is more likely to throw light on the origin and nature of the human race. The problem is, however, that not everything is perfectly clear in this field of research.

Life is indeed understood as a capacity and a process, but Life is also construed as a *force*, for the evolution of living beings is explained here by the fact that all of them possess certain properties that determine the emergence of a new law of nature, the law of natural selection, that describes the behavior of a new force of nature. Since the law that governs the evolutionary process concerns living beings alone and cannot be found anywhere else in nature, the force of nature whose behavior is described by the law of natural selection emerges as a sui generis force, termed "Life." Thus, Life in the modern sense turns out to be not only a process and a capacity, but also a force that is inherent in all living beings and keeps them alive.

There are some troubling implications that follow from this concept of Life. One of them concerns the fate of the Christian God. Based on Maynard Smith's definition of Life, for example, one cannot say that God is alive, even though God is "life" (*zoe*) according to the gospel, for the simple reason that God cannot reproduce and be subject to Darwinian evolution. Hence the eternal quarrel between Darwinians, who sometimes feel obliged to clarify in which sense

God may exist anyhow, and Christians, who sometimes go so far as to deny evolution itself to make their point. Another important implication that follows from the currently predominant view of Life is that natural selection should be understood as being a law of nature that originates from prior, more fundamental laws—physical, chemical, biological—but cannot be reduced to any of them. Hence the mystery surrounding Life's origin and all issues related to so-called emergent properties. Yet another implication, less obvious but no less important, partly explains why evolutionary biologists are prone to downplaying the notion that Life is a force of nature. The problem that they encounter is that this force appears to be radically different from all other forces conceptualized by modern science, not only because it belongs to living beings alone, but also because it has some special traits that make it somewhat suspect from a scientific viewpoint. For instance, it presupposes the existence of final causes.

In part I, the most general features of the Darwinian science of Life will be examined, and close attention will be paid to the confusion between nomological and axiological statements that characterizes this science. As will be shown, the naturalistic fallacy (i.e., the identification of the science of what is alive with the science of what is good) is not the result of some blatantly inadequate interpretation of the doctrine. Rather, it stems from its basic assumptions. Darwin's theory entails that Life is a teleological and/or teleonomic process. Therefore, Life always pursues goals, which in turn become the principles of what I call the "morals of Life." The consequence is the return to a premodern style of thinking that finds expression in two key ideas: first, Life establishes a natural-and-moral *hierarchy* among living beings; second, Life issues natural-and-moral *orders*.

In part II, the focus will be on the biopolitical implementation of the morals of Life. If biopolitics (or biopower) in the broadest sense means the government of living beings, biopolitics today should be thought of, more precisely, as the government of *modular* living beings. The modularity of living beings is a central tenet of modern evolutionary theory, just as the modularity of human beings is a central tenet of the present-day biopolitical technologies of power, which are not geared toward the *normalization* of individuals any longer but rather aim for the *optimization* of deindividualized behavioral patterns. This shift from normalization to optimization marks the beginning of a new biopolitical era: population thinking in biology is paralleled by population management that draws inspiration from research in evolutionary theory, economic science, and behavioral psychology.

In the conclusion, I will examine some bioethical dilemmas that arise as a result of the cryptometaphysical disjunction between Life and the living. In that regard, what is worth noting is not only the conceptual and logical structure of such dilemmas, but also the bewilderment of both health professionals and the public when they face them. This is a sign that the observers, whether professionals or laypeople, share with the observed the same assumptions about Life that cause those dilemmas to appear in the first place. In the introduction, I will start to discuss some of these assumptions by summarizing my previous research on the topic.[8]

Several issues addressed in this book had already been tackled by Michel Foucault in his pathbreaking and widely known works on biopolitics. Here, emphasis will be placed on what Foucault did *not* see: the importance of evolutionary theory for recent developments in biopolitics; the metamorphosis of an old biopolitical regime centered on social normalization into a new ethopolitical regime centered on behavioral optimization; the emergence of bioethical problems that are closely connected with the biological abstraction of Life from any medium, the human body included.

Earlier, incomplete drafts of some chapters of this book have been published in Italian journals. I have revised them for the present edition to such an extent that it now seems unnecessary to say where those drafts originally appeared. The same holds for the equally unsatisfactory draft of chapter 4, which had already been published in English.

The book begins and ends with some remarks about the COVID-19 pandemic, but let me make clear that this is not a book *about* the pandemic and the ensuing crisis. In my view, recent events are just the latest symptom of a wider biopolitical reconfiguration of Western societies.

Introduction: The Morals of Life: On Metaphysics

Shortly after the COVID-19 pandemic hit the Western world, a debate arose over the policies adopted by governments to avoid the spread of contagion and high rates of hospitalization. Lockdowns were particularly subject to criticism, as they seemed to put people's rights in jeopardy. While a large majority of the public and the intellectual elite welcomed emergency measures taken by nearly all the Western governments, a small minority disapproved of them, often in a very harsh tone, claiming that the pandemic threat was being overstated. The same political polarization reappeared some time later, once vaccination campaigns had been launched: many breathed a sigh of relief, whereas others frowned on vaccines, arguing that there was too little evidence for their efficacity and the absence of severe adverse reactions. Interestingly, despite such major disagreements, people on both sides of the dispute agreed on one point: the natural, primordial fear of death that all human beings have since birth was key to understanding what was going on. The majority thought that this fear explained and justified their own compliance with the unprecedented limitation of rights imposed by public authorities. The minority, on the other hand, thought that governments and international organizations, possibly influenced by private-sector entities, were deliberately playing on this fear to introduce arbitrary restrictions on freedom. Both the majority and the minority thus shared the view that humans are inherently terrorized by the prospect of death, and that for this reason alone, they can accept having their basic rights and freedoms drastically compromised. If something like that happened during the pandemic crisis, this was because people felt the need to protect themselves and their loved ones against the imminent danger of death, regardless of whether such danger was real or utterly exaggerated. In the following, I will complicate the picture and frame the whole question differently.

To start with, it is worth paying some attention to a slightly more sophisticated version of the aforementioned view, which is known as "terror management theory"—first proposed by the anthropologist Ernest Becker in the 1970s and then refined and further developed by a team of social psychologists in the 1990s.[1] Briefly put, the theory says that, following Darwin's teachings, it must be assumed that "all living beings share a biological predisposition toward self-preservation, because such a tendency facilitates staying alive long enough to reproduce and pass one's genes on to future generation."[2] Unfortunately, however, in the course of evolution, the human species has developed too high a degree of consciousness, thus becoming aware of the inevitability of death. To counter this biological reality and the maladaptive dread that it engenders, cultural beliefs have been created, allowing humans to think of themselves as immortal:

> What saves us is culture. Cultures provide ways to view the world—worldviews—that "solve" the existential crisis engendered by the awareness of death. Cultural worldviews consist of humanly constructed beliefs about the nature of reality that are shared by individuals in a group that function to mitigate the horror and blunt the dread caused by knowledge of the reality of the human condition, that we all die.[3]

Based on these premises, not only cultural worldviews but also a vast range of mental disorders and disturbed behaviors can be better understood, according to terror management theorists. Unsurprisingly, the theory applies to the pandemic crisis as well:

"Regardless of whether one consciously believes that the virus is a major threat to life or only a minor inconvenience, fear of death plays an important role in driving one's attitudes and behavior related to the virus."[4] For instance, to assuage death anxiety, some have trivialized the virus and minimized its threat, arguing that it was not as contagious or lethal as experts claimed it to be, or even comparing it to common illnesses such as the seasonal flu. Others, on the contrary, "have engaged in some form of social distancing, increased sanitation practices such as hand washing and cleaning surfaces, wore masks in public places, and done other things to stay healthy."[5] For terror management theorists, both attitudes attest that fear of death is paramount and always affects human behavior in one way or another. The problem remains, however, as to what the phrase "fear of death" means exactly. In the end, it all depends on how we understand death and its consequences. And interestingly, by looking at this problem from a historical and anthropological point of view, it appears that the meaning of death and the related emotional responses that death elicits vary across human cultures and over time.[6]

According to the historian Philippe Ariès, Western attitudes toward death have gone through several changes throughout the centuries. In the past, "the spectacle of the dead, whose bones were always being brought up to the surface of the cemeteries, as was the skull in *Hamlet*, made no more impression upon the living than did the idea of their own death. They were familiar with the dead as they were familiarized with the idea of their own death."[7] In our time, as Ariès emphasizes, things have changed:

> We have witnessed a brutal revolution in traditional ideas and feelings, a revolution so brutal that social observers have not failed to be struck by it. It is really an absolutely unheard-of phenomenon. Death, so omnipresent in the past that it was familiar, would be effaced, would disappear. It would become shameful and forbidden.[8]

Thus, based on Ariès's wide-ranging investigation, it is doubtful that an overwhelming fear of death may be deemed a universal, natural, and primordial feeling that affects the human condition always and everywhere in the same manner, as terror management theorists maintain. Rather, it becomes apparent that human attitudes toward death vary greatly depending on the historical period and the overall cultural background (not to mention personal dispositions). To assume that cultural beliefs operate only on a surface level—that they serve to mitigate a congenital and inescapable horror of death—is mere speculation.

The same point can be made from an anthropological point of view. According to Robert Hertz, the construal of death, as well as the emotional responses to the prospect of one's death or the deaths of others, are contingent on the way that a certain culture frames the bare fact of death. Perhaps it would be even more accurate to say that there are no bare facts here, or that the fact of death is always theory-laden. For Hertz, the theory in question amounts to the social or "collective representation" of death, which is part of a broader cultural worldview:

> We all believe we know what death is because it is a familiar event and one that arouses intense emotion. It seems both ridiculous and sacrilegious to question the value of this innate knowledge and to wish to apply reason to a subject where only the heart is competent.
>
> Yet questions arise in connection with death which cannot be answered by the heart because the heart is unaware of them. Even for the biologist death is not a simple and obvious fact; it is a problem to be scientifically investigated. But where a human being is concerned the physiological phenomena are not the whole of death. To the organic event is added a complex mass of beliefs, emotions and activities which give it its distinctive character. . . .
>
> Thus death has a specific meaning for the social consciousness; it is the object of a collective representation. This representation is neither simple nor unchangeable: it calls for an analysis of its elements as well as a search for its origin.[9]

Leaving aside the somewhat obsolete sociological (i.e., Durkheimian) language, Hertz's reflections clear up misconceptions about the problem with which we are confronted: death is not a simple and obvious fact, to which humans react in a simple and obvious manner, but rather a culture-laden phenomenon—"even for the biologist." It follows that much remains unexplained regarding the COVID-19 crisis.

Everybody seems to agree that fear of death played a crucial role during the pandemic and was instrumental in persuading most people to give up some of their basic rights for a while. But fear of death does not explain anything in the final analysis unless we clarify what we mean by death, and therefore what we truly fear. It is not an easy task to clarify this, since the perception of death changes considerably according to cultural beliefs and values. Thus, the questions become: What are our own cultural beliefs and values? And why, drawing on our cultural assumptions, do we end up considering death as something terrifying, "shameful and forbidden," as Ariès put it?

Here, I will contend that the reason for this has little to do with our alleged natural fear of extinction and much more with another fear, which I call "fear of Life itself": the fear that we are not living up to the demands of Life—whence our irrepressible feelings of fragility, defectiveness, and precariousness. In my view, this fear of Life hides behind what we label as our innate fear of death. I am aware that this thesis may sound highly counterintuitive, if not foolish. The aim of this book is to make it less incomprehensible and dubious than it may seem at first glance and to show that it can help us to understand, at least partly, what is going on these days.

But, to begin with, what is going on? It is no mystery that the coronavirus crisis has a twofold nature, biological and biopolitical. And to examine all the implications of such a biological-biopolitical crisis, we must start from an in-depth analysis of what we mean by Life—the life of which biology speaks, the life of which biopolitics should take care. Although it may seem that this move leads us astray, it actually allows us to cast light on the origin of several problems that we encounter today. If Western societies are now going through a process of momentous transformation, as the coronavirus crisis attests, this is because the ground for this change had been prepared over a long time.

As I have argued elsewhere, late modernity saw the emergence of a new metaphysical paradigm centered on a previously unknown concept that philosophers and scientists contributed to fashioning in the nineteenth and twentieth centuries: the concept of Life.[10] This concept is an abstraction invisibly related to another crucial abstraction of modern thought—the free will of the autonomous man. Based on the abstraction of Life as such, it became possible to establish a science of Life.

Although the word "biology" had already appeared a couple of times in eighteenth-century texts, it came into its modern usage at the beginning of the nineteenth century, when the new science or doctrine of Life—*Biologie* or *Lebenslehre*, as Gottfried Reinhold Treviranus originally called it—was born. This science, unlike previous naturalistic investigations, focused not only on the various forms and features of the living but also on the conditions and laws under which Life as such manifests itself in the world. In the mid-nineteenth century, Darwin established the framework within which debates about Life have been conducted ever since. Still today, when the question of Life comes up, one can only be for or against Darwin, who has grown into a *summa auctoritas* of science and may be considered the Aristotle of modern times.[11]

In the history of Western culture, the birth of biology and the rise of Darwinism mark a point of no return, not only because the life sciences have become essential to our understanding of nature, but also because Darwinism has extended its influence beyond the field of biology, affecting the evolution of other sciences, from sociology to psychiatry, from economics to political science, not to mention various branches of philosophy and cognitive science. In light of these developments, it is no surprise that biology and Darwinism have also influenced the way in which the government of human beings has been conceived and put into practice. Eugenics, social Darwinism, and sociobiology are some examples of the biologistic turn in the social sciences that has contributed to the biopolitical reshaping of late modern societies.

Philosophers and social scientists have often discussed the merits and faults of the Darwinian approach to nature and human beings. John Dewey was among the first to point out that "in laying hands upon the sacred ark of absolute permanency, in treating the forms that had been regarded as types of fixity and perfection as originating and passing away, the *Origin of Species* introduced a mode of thinking that in the end was bound to transform the logic of knowledge, and hence the treatment of morals, politics, and religion."[12] Over the last century, however, many have wondered whether this transformation had been carried out in the name of science or in the name of metaphysics. Karl Popper famously stated in his autobiography that Darwin's theory is more like a "metaphysical research programme."[13] Imre Lakatos had similar doubts.[14] More recently, Jerry Fodor and Massimo Piattelli-Palmarini have placed emphasis on the "intensional fallacy" that vitiates neo-Darwinian speculations, stressing that within and outside the field of biological research,

> neo-Darwinism is taken as axiomatic, it goes literally unquestioned. A view that looks to contradict it, either directly or by implication, is *ipso facto* rejected, however plausible it may otherwise seem. Entire departments, journals and research centers now work on this principle. In consequence, social Darwinism thrives, as do epistemological Darwinism, psychological Darwinism,

> evolutionary ethics—and even, heaven help us, evolutionary aesthetics. If you seek their monuments, look in the science section of your daily paper.[15]

One can hardly disagree with Fodor and Piattelli-Palmarini on this point: neo-Darwinism looks like a dogmatic creed today. Yet, besides denouncing the discrepancy between dogma and science, we should ask ourselves why the Darwinian, and then the neo-Darwinian, interpretation of evolutionary processes is considered axiomatic by most people, whether scientists or laypeople. In my view, the reason is that the theory of natural selection cannot be uncoupled from what I call the metaphysics of Life. In a nutshell, this amounts to saying that such a theory reinforces and gives full expression to a conception of Life that has become widespread in the late modern period, according to which Life is not only a natural process or force but also a moral value. The modern metaphysics of Life consists precisely in this confusion between the fact and the value of Life, or between the nomological and the axiological characterizations of Life.[16] Darwin's theory and Darwinism more generally are the most telling and influential of the many versions of this metaphysics that have appeared over the last two centuries. No doubt, they are also the most important from the point of view of political history.

What is Life? In modern times, the question has become crucial for many reasons, which are not just related to science and philosophy. Michel Foucault was the first to highlight that the very life of human beings fell within the scope of public policy in the eighteenth and nineteenth centuries. This was the moment when biopower (in Foucault's sense of the word) gained impetus. Long story short, states were charged with the task of monitoring the lives of citizens through disciplinary and regulatory apparatuses. Foucault speaks of an "anatomo-politics" of the human body and of a "bio-politics" of the human population, which represent the two sides of modern biopower. The former was aimed at subjugating individual bodies, making them docile and exploitable (the school, the army, the jail, and the asylum were the environments in which this kind of dressage was typically pursued). The latter targeted the body of the species and the related biological processes (birth, health, hygiene, longevity, reproduction, and others) for the purpose of achieving "a calculated management of life."[17]

With the passage of time and the growth of state apparatuses, particularly noticeable during the late nineteenth and early twentieth centuries, biopower further widened the range of interventions and improved its action with the help of new scientific tools provided by medical and biological research. The science of epidemiology, for instance, went through various stages of development during this period of time and became instrumental in the regulatory control of populations. As Foucault points out, it was then that biopolitics began to morph into thanatopolitics as a result of too strict an interpenetration between the language of politics and the language of biology. The worst forms of racism appeared at that point:

> Racism took shape at this point (racism in its modern, "biologizing," statist form): it was then that a whole politics of settlement (*peuplement*), family, marriage, education, social hierarchization, and property, accompanied by a long series of permanent interventions at the level of the body, conduct, health, and everyday life, received their colour and their justification from the mythical concern with protecting the purity of the blood and ensuring the triumph of the race. Nazism was doubtless the most cunning and the most naïve (and the former because of the latter) combination of the fantasies of blood and the paroxysms of a disciplinary power. A eugenic ordering of society, with all that implied in the way of extension and intensification of micro-powers, in the guise of an unrestricted state control, was accompanied by the oneiric exaltation of a superior blood; the latter implied the systematic genocide of others and the risk of exposing oneself to a total sacrifice.[18]

As Foucault explains, Nazism epitomizes the biologistic drift of late modern biopolitics. Curiously, however, neither Foucault nor his epigones have ever paid much attention to the Nazis' science of Life. As others have recalled and I myself have emphasized in my previous contribution to this topic, social Darwinism was a crucial ingredient of Nazism.[19] In some sense, the most relevant aspect of Nazism from a biopolitical point of view lies in the fact that the Nazis took the twofold semantics of modern Life—epistemic and moral—very seriously. The Nazis were not only the cruel students of the *Lebensunwertes Leben* in the concentration camps but also the high priests of the Aryan race, in which Life was supposed to have found its strongest expression. For them, Life was not just a natural process or force. It was also, and above all, an axiological principle. Social Darwinism had taught them this lesson.

Until not so long ago, none of this would have been possible, for Life did not exist. As Foucault explains in *The Order of Things*, all that existed were living beings, not Life as such, or Life abstracted away from the living:

> Historians want to write histories of biology in the eighteenth century; but they do not realize that biology did not exist then, and that the pattern of knowledge that has been familiar to us for a hundred and fifty years is not valid for a previous period. And that, if biology was unknown, there was a very simple reason for it: that life itself did not exist. All that existed was living beings.[20]

But how is Life to be understood once it has been abstracted from living beings? Surely we cannot content ourselves with a definition like this: Life is the opposite of death. This tautology leads nowhere, as Samuel Coleridge remarks.

> The physiologist has luminously explained Y plus X by informing us that it is a somewhat that is the antithesis of Y minus X; and if we ask, what then is Y-X?

> the answer is, the antithesis of Y+X . . . The definitions themselves will best illustrate our meaning. I will begin with that given by Bichat. "Life is the sum of all the functions by which death is resisted," in which I have in vain endeavored to discover any other meaning than that life consists in being able to live.[21]

Thus, the question comes up again: What is Life? In short, we may say that Life in the modern sense is a process and, more important, a *force*. This is where Darwin and Darwinism enter the picture. Life is a force in a Darwinian perspective because it follows a law of nature discovered by science, just like gravitation. According to Newton, we can look at gravitation as a natural force because we can think of a scientific law that describes and predicts the attraction between all physical bodies. Likewise, according to Darwin, we can look at Life as a natural force because we can think of a scientific law that describes and predicts the behavior of all living beings. The name of this law is "natural selection."

It is not easy to explain the theory of natural selection, for it lends itself to two different interpretations that are, more often than not, joined together in Darwin's and his followers' works. According to one interpretation, the "struggle for existence" favors those living beings (whether genes, organisms, or groups) that develop adaptive traits and fit in better and better with the environment in which they find themselves living. According to the other, natural selection favors those living beings that increase their capacity to go through further selection by inventing ever-new ways in which Life can be breathed into the living, and by creating, as it were, the related ecological niches.[22]

The first, adaptation-oriented interpretation conveys a trivial understanding of natural selection. Here, as Fodor and Piattelli-Palmarini say, natural selection seems to be carried out by an intentional system that can discriminate between various options and see which trait is adaptive and which not—and therefore, which trait must be selected for a certain function in a certain environment and which not. It is quite evident that this interpretation is vitiated by teleological reasoning:

> If genes were intentional systems, or if there were a Mother Nature who selects with ends in view, then there would be a matter of fact about which traits they select for and which traits are merely coextensive with the ones they select for. That's the good news. The bad news is that, unlike natural selection, Mother Nature is a fiction; and fictions can't select things, however hard they try.[23]

Besides the fact that Mother Nature is a fiction and fictions do not exist in real life, there is another reason why the adaptation-oriented interpretation of natural selection cannot be deemed valid. As a universal law of nature, natural selection should apply to *all* cases of evolutionary adaptations. Yet environments and mechanisms of adaptation change from place to place, and one cannot make claims that generalize over the totality of situations that follow one another in natural history:

> There's a story about how spiders catch flies to eat, and there's a story about how oak trees distribute their seeds, but the two have little or nothing in common; there aren't, as a philosopher might put it, laws—or even reliable empirical generalizations—about their mechanisms of adaptation or the structure of their niches. Some of them work in one way, others work in quite different ways, and no two are likely to work in much the same way.[24]

This is only half the story, however, for there is another, more sophisticated interpretation of Darwin's theory, the selection-oriented interpretation, which is less exposed to Fodor and Piattelli-Palmarini's objections. This interpretation says that Mother Nature selects for the purpose of further selection.[25] I will call this version of Darwin's theory "selectionism," even though it is often labeled "adaptationism" in the literature and lays the groundwork for the "adaptationist research programme," on which I will expatiate later. In this view, reproductive success—rather than adaptation—is the primary aim of natural selection because reproductive value maximization increases diversification among living beings, broadening the range of possibilities and choices for future, ever more complex selection.

The result of this process is a spontaneous, unstoppable increase in the degrees of complexity and organization of the biosphere. By means of natural selection, Life grows in the world, pushing life forms to multiply their chances of survival by multiplying their traits and structures. Clearly, the environment continues to play a role in this scenario, in that it withstands the pressure toward reproductive value maximization and thus favors some life forms while condemning others to be eclipsed by "the fittest," those that reach higher rates of reproduction. The point is, however, that the sieving action of the environment can be moved to the background here because there is no need—for either Mother Nature or the biologist—to know ex ante what kind of resistance the environment will exert each time. Given the universal, constant pressure toward reproductive value maximization, some life forms will survive while others will go extinct anyhow. In this perspective, we need not think that natural selection *sees* anything, or that it intentionally *selects for* this or that adaptive trait.[26] All we can say ex post is that new ecological niches (not to be confused with habitats) emerge one after the other and inflate the biosphere under pressure from natural selection, which does nothing but broaden the gamut of selective choices from one generation to the next.[27] What is commonly termed "adaptation" follows as a consequence of this underlying process. In the first edition of *On the Origin of Species*, Darwin illustrated this idea as follows:

> In looking at Nature, it is most necessary to keep the foregoing considerations always in mind—never to forget that every single organic being around us may be said to be striving to the utmost to increase in numbers; that each lives by a struggle at some period of its life; that heavy destruction inevitably falls either on the young or old, during each generation or at recurrent intervals. Lighten

> any check, mitigate the destruction ever so little, and the number of the species will almost instantaneously increase to any amount. The face of Nature may be compared to a yielding surface, with ten thousand sharp wedges packed close together and driven inwards by incessant blows, sometimes one wedge being struck, and then another with greater force.[28]

The last word of this passage, "force," is crucial for two reasons. First, according to selectionism, Life is not just the opposite of death because Life, understood as a force that transcends living beings, does not simply denote the capacity to resist death for a while. Rather, Life is that which *survives and conquers death.* Second, based on the same assumption, selectionism can avoid teleological reasoning. Here, the purpose of Life turns out to be Life itself, which aims for self-reproduction. Life therefore can be considered an autotelic force. In this regard, one may speak, with Jacques Monod, of a "teleonomic" rather than teleological scheme of things.

Following this line of reasoning, two further claims can be made. First, the autotelic purposiveness without any (external) purpose is the most general and distinctive feature of Life as such, abstracted from living beings. Second, this feature is so general that it can explain all episodes of speciation that have occurred in the history of evolution. In this view, all living beings pursue a purpose that is unrelated to the varying environmental conditions: namely, reproductive value maximization.[29] Thus, teleonomy turns into a universal law of nature that identifies Life with a force of nature that causes new species to arise in the course of natural history.

In the first part of this book, I will examine in more detail this version of Darwin's theory and explain why it lies at the bottom of the most widely accepted understanding of biological processes. In the second part, I will show how the neo-Darwinian approach contributes to shaping the new grammar of knowledge that undergirds today's biopolitical practices. To conclude this introduction, I will roughly outline some implications of the neo-Darwinian conception of nature.

One of the reasons why the neo-Darwinian view of Life is so important is that it brings into the open the metaphysical background against which the modern view of nature and human beings has developed. According to neo-Darwinism, or selectionism, Life can be regarded as a natural power that animates living beings, in that the latter can be seen as the "selfish" agents that endeavor, with greater or lesser success, to maximize reproduction rates.[30] Based on this premise, neo-Darwinism seems able to provide a nomological explanation for evolution. The problem is, however, that the selection-oriented interpretation of Darwin's theory is objectionable from the point of view of rigorous science, not only because there is no way to test it, but also and above all because the nomological and the axiological regimes of enunciation become confused here.

To clarify this, let us take one step at a time. The first thing to note is that what looks like a natural necessity from the biologists' perspective turns into a natural imperative for the living—the imperative of maximizing reproductive success, which weighs on all creatures. In the end, it is not clear why we should hold on to this idea. As Niles Eldredge remarks, we may well "see reproduction as a physiological luxury rather than an imperative that is necessary for that fox to go on living."[31] Said in a slightly different way, the thesis that reproductive value maximization is a natural necessity is in the last instance an a priori assumption, for it is not suggested unequivocally by observational data. Quite the contrary, in fact—it is this assumption that pushes scientists to ascribe more value to certain data, leaving other information aside.

As a result, some observational data, such as reproduction rates, are set apart and judged more relevant than others in the belief that reproductive value maximization is the most pressing need for all living beings. In line with this approach, reproductive value maximization is thought of as the ultimate purpose—or the "primary project," as Monod has it—of all living beings.[32] Absent this purpose or project, evolution by means of natural selection would stop, and Life would vanish from the face of the earth.[33] The reasoning seems faultless to many biologists, but this is where problems start.

At this point, the theory of natural selection reveals itself to be based on an amphiboly: reproductive value maximization is deemed a natural necessity and, at the same time, a natural finality that governs the life of living beings—a finality arising from within the living (reproduction, survival) rather than from the outside (adaptation with ends in view), but a finality nevertheless. The amphiboly is unavoidable, and so is one of its logical consequences: if an increase in reproduction rates can be understood as the abstract telos that living beings must pursue always and everywhere, finding ever-new ways to comply, then reproductive success can be understood as a criterion for value-laden statements that compare living beings endowed with different evolutionary traits, and that classify them as being better or worse. In this manner, nomological statements become indistinguishable from axiological pronouncements: those who survive appear to be those who follow more closely the law/imperative of Life, the law/imperative of reproductive success, which turns into the basic parameter for the Darwinian science/morals of Life.

Once this conclusion has been drawn, we find ourselves in the vicinity of another concept which, according to Georges Canguilhem, has an axiological meaning: health. "*Valere*, from which value derives, means to be in good health in Latin. Health is a way of tackling existence as one feels that one is not only possessor or bearer but also, if necessary, creator of value, establisher of vital norms."[34] The same can be said of Life in the Darwinian sense. Here too, to be alive means to be the bearer and creator of values, that are restated and reinforced throughout evolution: reproduction, survival.

In light of this, it seems difficult to draw a line between science and metaphysics, nomology and axiology, when we look at natural history from the viewpoint of Darwinism, neo-Darwinism, or ultra-Darwinism. Within this framework, axiological and nomological principles are two sides of the same coin. Life is a natural necessity but also a moral imperative addressed to the living. In more abstract terms, the logic of Darwin's view of Life is the same that one can see at work in Immanuel Kant's moral reasoning. This parallelism makes the axiological implications of selectionism even more evident.

For Kant, the human will can be thought of as autonomous—and therefore we can think of ourselves as free human beings—if and only if the will is not conditioned by anything else. Inevitably, however, every act of volition in which our will finds expression is conditioned by something else. "I will something"; otherwise, my will remains silent. Thus, one's will is conditioned by that which one wills. This means that one's autonomy ("I will . . .") is just a formal, a priori requirement of all material acts of volition that are nonetheless doomed to transgress it on a regular basis ("I will . . . something"). The formal autonomy of one's pure will is continuously vitiated by the material heteronomy of all acts of volition. As Kant put it, the unconditioned will is the "fact of reason" because "I will" every time "I will something," but this "fact of reason" is not a fact, properly speaking, as it is always conditioned by the empirical goals that contaminate it. The outcome of this philosophical tour de force is as follows: the autonomous will takes the shape of a nomological necessity that is ascertained through critical reasoning; and yet this necessity is axiological in nature, in that it represents nothing more than a cogent imperative to be restated throughout one's existence (the categorical imperative, as opposed to the hypothetical imperatives in which it materializes).

The similarities with Darwin's argument about Life are apparent—in particular, the conflation between the nomological and axiological perspectives that paves the way to the extraction of a metaphysical surplus value, of a *natural-and-moral force* from the body of experience: the force of the will in one case, the force of Life in the other. Autonomy is the conceptual invariant that bridges the gap between natural and moral ontology. For Kant, the will is a force governed by a nomological necessity that prescribes the will's autonomy from all material acts of volition under the imperative of sanctity; for Darwin, Life is a force governed by a nomological necessity that prescribes its autonomy from all material living beings under the imperative of survival. No wonder, then, that here, sanctity means the survival of the autonomous will (the eternal life of the soul), while survival means the sanctity of Life as such (what the living value most). The two imperatives are ultimately the two faces of the same medal, the metaphysics of autonomy, the metaphysics of modernity, based on the fungibility of moral and natural properties.

This new kind of metaphysics, which blurs the lines between nomology and axiology, gives rise to a particular kind of ontology—the "untamed ontology," as Foucault calls it, an *ontologie sauvage* that seems to revive some sort of archaic *pensée sauvage*, which blurs the lines between being and nonbeing.[35] If the will takes the form of an imperative, in fact, this is because the autonomous will never will enough and must reaffirm itself in ever-new ways throughout one's existence. The being of the autonomous will suffers from a lack of being. Likewise, if Life takes the form of an imperative, this is because it never lives enough and must reaffirm itself in ever-new ways throughout evolution. The being of Life itself suffers from a lack of being. Thus, modern people remain trapped between the two foci of the same ellipse, the metaphysics of autonomy, which condemns them to be the transitory and ever-more-intimidated figures of a will in search of content, on one side, and of Life's never-ending search for survival, on the other, as Foucault says:

> The experience of life is thus posited as the most general law of beings, the revelation of that primitive force on the basis of which they are; it functions as an untamed ontology [*ontologie sauvage*], one trying to express the indissociable being and non-being of all beings. But this ontology discloses not so much what gives beings their foundation as what bears them for an instant towards a precarious form and yet is already secretly sapping them from within in order to destroy them. In relation to life, beings are no more than transitory figures, and the being that they maintain, during the brief period of their existence, is no more than their presumption, their will to survive.[36]

I The Morals of Selection: On Biology

1 How Darwin Changed Philosophy

1.1 The Discovery of Life

To study Charles Darwin's contribution to philosophy, one should first look at his philosophical sources and then examine the differences between Darwin's own thought and these sources. Yet this task reveals itself to be highly problematic, for reasons that need to be made explicit from the outset: first, Darwin was not a scholar trained in philosophy or in the discipline that we ordinarily call by that name; second, and more important, it is extremely difficult to make a clear distinction between what is philosophical, properly speaking, and what is not. I mention this because Darwin certainly had his philosophical sources, but these sources were not philosophical in the conventional sense. Richard Owen and Joseph H. Green (who coined the phrase "the tree of life," which Darwin later appropriated), two English naturalists of the early nineteenth century with whom the young Charles came in contact, and Alexander von Humboldt, one of Charles's intellectual heroes, were not philosophers *stricto sensu* but they did not dislike engaging in what Darwin labels, in his *Notebooks*, "metaphysical speculations."[1] And the same can be said of other, more exotic thinkers to whom Darwin sometimes makes reference in his *Notebooks*, such as Carl Gustav Carus, a follower of Friedrich Schelling. Briefly put, the boundary between what we call "biology" (or the "life sciences") and the philosophy of nature (*Naturphilosophie* in German) was not well defined at the time. And even in our day, it is not a foregone conclusion that this boundary can be clearly outlined.

Going back to the question of how Darwin's thought affected the whole field of philosophy, the simplest and perhaps the truest answer reads as follows: with his *biology*, Darwin not only stated that Life exists, but also made this new thing, Life, somehow palpable to scientists by abstracting it from all its material embodiments (namely, living beings and life forms).[2] The reason why Darwin is so important for us, therefore, is not because he introduced the idea of evolution, of species transformation. Others before him, including his grandfather Erasmus Darwin, had already foreshadowed or overtly discussed this idea. But Charles was surely the first to conceive of Life as an object of scientific investigation and to understand it as the motor of evolution, as the force that causes species transformation by means of natural selection. Darwin was thus the one who discovered—or, if you wish, invented—the biological concept of Life. Still today, according to most biologists, "Life should be defined by the possession of those properties which are needed to ensure evolution by natural selection."[3]

With this idea in mind, Darwin was able to consolidate the field of research that takes the name "biology," the study of Life. As Jean Gayon has remarked, in fact, "the doctrine of natural selection" is not only one theory among the many that can unify the whole of biology; rather, it is, more profoundly, "the only possible theory that can fulfill such a task."[4] In other words, according to Gayon and others, Darwin's theory of natural selection is *the* theory of Life that delineates the contours of biological research. As Theodosius Dobzhansky put it years ago, "nothing in biology makes sense except in the light of evolution"—and there can be no doubt that by "evolution," Dobzhansky meant only and solely Darwinian evolution.[5] So, even though many have criticized and challenged the theory of natural selection over the years, Darwin's theory remains the cornerstone of today's biology. The modern evolutionary synthesis of the 1930s and 1940s introduced many remarkable novelties, to which I will return later in this discussion, but neo-Darwinism keeps sharing the same fundamental hypothesis or principle that lies at the heart of Darwin's original argument.

1.2 The Role of Ignorance

The two key concepts of Darwin's theory are selection and variation, upon which selection acts. Both concepts help to explain the "improvement" of life forms in the course of natural history, and both are critical to Darwin's view of Life, which he describes in magniloquent terms toward the end of his masterpiece.[6] "There is grandeur in this view of life."[7] What is perhaps less glorious is the role that ignorance plays in this theory of Life.

It is well known that Darwin loved comparing his theory to Newton's theory of gravitation. The parallelism is not without reason.[8] Newton's theory focuses on a natural force that is not the property or the attribute of certain physical bodies but rather the manifestation of a natural law that governs all of them. The gravitational force determines the movement of physical bodies differently to Aristotle's theory of the natural places, according to which some bodies go up (air and fire) and others down (water and earth) by virtue of their physical nature. Hence, it can be said that Newton's theory abstracts the gravitational force from the inherent qualities of the various physical bodies, which all follow the same universal law of attraction. As for Darwin's theory, it does the same with Life, which it abstracts from the species-specific traits of the living. For Darwin, natural selection rules over *all* living beings. For this to be possible, however, a certain amount of ignorance is needed. And this is where things get interesting.

As far as variation is concerned, Darwin famously contends in *The Origin of Species* that he knows neither its laws nor its causes. Ignorance affects the scientific observer in this case, and such ignorance is by no means fortuitous: in the end, it is only because we ignore the laws and causes of variation, and so are unable to trace them back to the species-specific traits of this or that living being, that we can begin to abstract Life as the universal *variability* of all living beings. From a Darwinian point of view, in short, biologists' ignorance of the mechanism of variation, however regrettable it may be, enables them to take a first step toward reaching a theory of Life that disconnects Life as such from the living, and that understands Life as a universal and potentially infinite plasticity of life forms. In that regard, it is perhaps also worth recalling that when Darwin sought to explain the mechanism of variation with the help of his unfortunate theory of "pangenesis," the result was that the very foundations of his theory of natural selection were shaken.[9] August Weismann was the one who restored them, shortly before the modern evolutionary synthesis took center stage in debates among theoretical biologists.

As far as selection is concerned, ignorance plays an even more overt role and, from being an invisible epistemic principle, it becomes ontological. In this case, it is no longer the scientific observer who is ignorant, but Life itself that has to confess its ignorance. Indeed, as Darwin often stresses, natural selection cannot but be a blind selection or, to be more precise, an "unconscious selection," for it ignores—and *must* ignore—the final result of its choices.[10] This is the reason why Darwin expands on the selection carried out by breeders, in which the winnowing of the best animals is partly "methodical" and partly "unconscious," highlighting the similarities between Life's natural selection and the breeder's unconscious selection. In a nutshell, the point here is that no such thing as natural selection would be conceivable if the latter could *see* in advance what the effects of its action will be in terms of better or worse adaptation of the living because in that case, one would tacitly reintroduce the idea of an intelligent designer and conceive of natural selection as an intentional system that guides the evolutionary process. For Darwin, by contrast, there is no intentional system or intelligent power behind evolution. Natural selection is ignorant of what it does.

With such a heavy emphasis placed on ignorance, that may possibly explain Darwin's own doubts about his "one long argument" and enduring perplexities about the publication of *The Origin of Species*, which was postponed several times, Darwin's theory nonetheless reaches a strategic goal: the abstraction of Life as such, an abstraction that consolidates the field of biology around two complementary ideas.[11] The first is the idea that Life is the secret *unity* from which all living beings and life forms flow, spreading all over the natural world—in Michel Foucault's words, the idea that Life is "the great, mysterious, invisible focal unity, from which the multiple seems to derive, as though by ceaseless dispersion."[12] Second, there is the idea that Life is a secret *force* that must be untangled and abstracted away from the living—a force, as Foucault writes, that is "inaccessible in its essence, apprehendable only in the efforts it makes here and there to manifest and maintain itself."[13]

1.3 The Chain of Being

Darwin's theory is like a drama in two acts, a drama that should be termed *metaphysical*, if only because the conceptual pillars that it eventually took down were themselves metaphysical. For the time being, I will leave aside the controversial question of whether Darwin's theory is *intrinsically* metaphysical rather than scientific. Whatever the case, the conceptual principles that the theory of natural selection not only calls into question but utterly wipes out are of a metaphysical nature: the principle of the great chain of being and the principle of *conservatio vitae*. In general terms, we may begin by saying that, with Darwin, one moves from a static conception of the natural world to a dynamic conception of natural processes, which are no longer caged in a preestablished and unalterable order of things. Darwin was not the first to advance this idea; others had discussed it before him (Schelling, for instance, who exerted a heavy influence on German *Naturphilosophie* and Romantic biology). But Darwin is the one who made the most of it.

The principle of the great chain of being says that the cosmos is a closed whole, with a top and a bottom, in which all creatures are ordered hierarchically, with no empty spaces in between them. This principle can be split into two subprinciples: the principle of graduality, according to which, from the most imperfect of beings, one progressively climbs up to the supreme and most perfect being; and the principle of plenitude, according to which there are no gaps between one level of reality and the next. In this view, all the natural forms and beings that populate the world are distributed on some sort of ontological pyramid, the *scala naturae*, which represents the Form of all forms, so to speak, and which suffers from *horror vacui*. For centuries, thinkers had elaborated on this notion of a continuous and hierarchically structured cosmos. Giovanni Pico della Mirandola, for example, ends his fresco of the Creation with the following words: "All space was already filled; all things had been distributed in the highest, the middle and the lowest orders."[14] The natural world, in Pico's view, is enclosed in the "universal order," *in universi serie*. Three centuries later, the same image returns in the words of Johann Gottfried Herder:

> When the door of creation was shut, the forms of organization already chosen remained as appointed ways and gates, by which the inferior powers might in future raise and improve themselves, within the limits of nature. New forms arise no more: but our powers are continually varying in their progress through

> those that exist, and what is termed organization is properly nothing more than their conductor to a higher state. . . . What the all-vivifying calls into life, lives, whatever acts, acts eternally in his eternal whole . . . It was not our reason that fashioned the body, but the finger of God, organic powers.[15]

Within this metaphysical framework, one cannot tell natural forces and natural forms apart: the former express themselves in the latter, thereby expressing themselves in a language, a *verbum*, that is the language of God. It goes without saying that Darwin shatters this framework, for five reasons at least.

First, Darwin maintains that Life is a propulsive force that never coincides with the life forms into which it pours itself; for him, natural selection *causes* the transmutation and evolution of species, of organic forms, without a pause; it is therefore related to a force that never exhausts itself in the forms that arise out of evolution. Second, ignorance rather than intelligence—whether or not of a divine nature—plays a paramount role in Darwin's theory and functions as an explanatory principle, for natural selection is a blind and unconscious selection; it does not know in advance what direction it will take. Third, we cannot think of any preestablished and unalterable order of things any longer; at this point, a static conception of the natural cosmos gives way to a dynamic view of vital phenomena. Fourth, in this perspective, there is always a gap to be filled, a new evolutionary possibility to be explored, because natural forms are not enclosed *in universi serie*; hence, we can bid farewell to the old metaphysical principle of plenitude (while holding firm to the principle of graduality: *Natura non facit saltum*, as Darwin often remarks). Fifth, following Darwin, we cannot say that absolute perfection can be found in nature; on the contrary, "natural selection will not necessarily lead to absolute perfection; nor, as far as we can judge by our limited faculties, can absolute perfection be everywhere predicated."[16]

1.4 Life as a Quantity

Given that natural selection does not know in advance how living beings will be transformed by its action and what kind of *formal* and *qualitative* features they will acquire, it remains to be seen what criteria natural selection follows when making its choices. Darwin's solution to this problem is ingenious: natural selection can be understood as being blind and unconscious only if it chooses, every time, to keep choosing *simpliciter*. According to Darwin, in other words, natural selection does not favor any evolutionary direction at the expense of another on the grounds of *qualitative* criteria and with a view to the species-specific traits that the living may have in the future, but rather on the grounds of purely *quantitative* criteria and with a view to the propagation of Life that life forms will be able to achieve.

Therefore, if natural selection is blind and evolution does not proceed with ends in view, this is because the sole end of evolution is the continuation of evolution itself. For evolution to extend itself indefinitely, the only requirement is that the living reproduce as much as possible. Natural selection is the name of this maximization process, which results in a perceived "improvement" of species: the best living beings are those that replicate the most, and the *more* they replicate, the *better* they become.

If we really agree with Darwin that natural selection is blind, we cannot add anything to this point. According to the hypothesis of natural selection, Life is just a force. And in principle, it seems perfectly possible to measure and quantify this force in terms of reproduction rates, or uneven levels of replication among different life forms. In this manner, biology starts qualifying as a modern natural science, since it will now require that biologists make accurate measurements and do mathematical calculations.

The possibility of biological measurements and calculation is one of the undeniable benefits of Darwin's theory compared to previous accounts of vital phenomena. The modern evolutionary synthesis will take great advantage of this possibility, and neo-Darwinism in general will make of it a source of scientific legitimacy. In Darwin's *Notebooks*, we find the traces of the difficulties that he had to overcome to reach this result and open the door to the future. There is, for instance, an extremely interesting note on Johannes Müller, a German naturalist and a contemporary of Darwin's, in which we can sense the effort that Darwin was making to get rid of one of his deep-rooted beliefs: "with respect to the non-development of Mollusca, which I have sometimes speculated might be owing to absolute quantity of vitality in the world: the production of vitality, as argued by Müller from propagation of infinite numbers of individuals from one of adverse."[17]

What is Darwin talking about here? Why is he opposing the idea of a "production of vitality" to the idea of an "absolute quantity of vitality in the world"? And why had he thought until then that the quantity of vitality, the totality of the vital force in the universe, was constant and should remain so? We do not have to go too far to find the reason: this was what everybody at the time took for granted. In a static and closed cosmos, in a universe nailed to an eternal order, the quantity of vitality (and of living beings) that the world contains is destined to remain unchanged, or "absolute."

So unquestioned was this assumption that even the naturalists who had ventured that life (or vitality) might be quantified and abstracted from the species-specific qualifications of living beings could not imagine taking this abstraction to its extreme yet logical consequences. Georges-Louis Leclerc de Buffon, for example, who had come to the idea that vital phenomena might be quantified well before Darwin, never dared to draw from this premise the conclusion that the "quantity of organic matter" in the universe could increase (or decrease) because he was aware that such an increase (or decrease) would entail altering, and ultimately destroying, the "form and consistence" of Nature itself:

In earth, air, and water, then, there exists a certain quantity of organic matter which cannot be destroyed, but which is constantly assimilated in a certain number of moulds, that are perpetually undergoing destruction and renewal: these moulds, or rather individuals, tho' varying in number in every species, are nevertheless always the same, that is, proportioned to the quantity of living matter; and this appears to be absolutely the case, for if there were any redundance of this matter, or if it were not at all times fully occupied by the individuals of the species which exist, it would, most assuredly, form itself into new species. . . . It is by this constant combination, and invariable proportion, that Nature preserves her form and consistence.[18]

Thus, Buffon rules out the possibility that new natural species arise out of a surplus of organic particles. What Darwin will do—and not without effort—is the exact opposite: he will assume that the quantity of vitality "in the World" can change, that there can be a "production of vitality," and therefore new species can emerge from Nature. Nobody, not even Thomas Malthus, had gone that far, for fear of disfiguring the face of Nature. For Malthus, the growth of populations (or the "principle of population") was far from implying a violation of the "great restrictive law" that governs the natural world.[19]

1.5 Disillusionment

In light of all this, Darwin's major innovations in the field of philosophy *lato sensu* can be summarized as follows: he abolished the principle of plenitude and introduced the idea that Life is a measurable force that pertains to all life forms but does not coincide with any of them; he theorized that not only the attributes of the living change in the course of evolution, but also the quantity of Life on earth is subject to change; he posited that perfection and eternity are not of this world.

Needless to say, there are many more aspects on which one should dwell to offer a complete picture of how Darwin changed philosophy. For instance, one should draw attention to the fact that the theory of natural selection disproves (or radically modifies) the very idea of natural species and paves the way for what has been called "population thinking" in the twentieth century.[20] Or one should discuss in detail all the problems concerning the living's adaptation to the environment. I will return to some of these problems later in this book.

Here, to conclude this chapter, I would rather insist on something else. Given that Darwin exerted such a heavy impact on Western thought and philosophy, one can hardly deny that he belongs to the history of Western philosophy. This does not mean only that the study of *The Origin of Species* should become part of all curricula in philosophy. It means, first and foremost, that Darwin's works should always be read with an eye to their ultimate philosophical message. Unfortunately, this message is often misunderstood, especially by scientists, and even more by neo-Darwinian biologists, who maintain that modern science in general and Darwinism in particular have the task of disillusioning people—with a special emphasis on religious disillusionment.

What I would like to suggest is that, on closer inspection, things are more complicated. If Darwin has disillusioned human beings, this is because he plunged them into ignorance. And by "them," I mean not only the common people but also scientists. Perhaps, this is the most significant philosophical lesson that can be drawn from Darwin's thought, a lesson in humility. Today, biologists seem not very sensitive to this lesson. Following Darwin, for example, it has become customary among neo-Darwinian theorists to elevate chance—that is, random genetic mutations—to the rank of an explanatory principle.[21] Chance and necessity are both integral to the most influential versions of the theory of natural selection that have been discussed over the last decades. But how can chance (namely, the partial or total unpredictability of certain phenomena) be transformed into a key to understanding natural phenomena? To this utterly philosophical question, Darwin gave an answer that is radically different from the one preferred by his epigones. His answer made it plain that chance means ignorance:

> I have hitherto sometimes spoken as if the variations—so common and multiform with organic beings under domestication, and in a lesser degree with those under nature—were due to chance. This, of course, is a wholly incorrect expression, but it serves to acknowledge plainly our ignorance of the cause of each particular variation.[22]

2 How Darwinism Changed Science

2.1 Explanation

According to Darwinism, a natural process drives life forms to evolve by means of variation and preservation of traits, which gradually give birth to new species and lead to the extinction of others. "Natural selection" is the name of the law that governs this process. The question that I will tackle here is whether the theory of natural selection truly explains evolution. Karl Pearson, an early advocate of Darwin's theory, did not think so:

> The biologist looks to force, chemical constitution, molecular structure, for an *explanation*, where at best they can merely provide conceptual shorthand for descriptive purposes. It seems all the more necessary to emphasize and repeat this important distinction, because the failure to grasp it has been made the ground for what is really a metaphysical attack on the Darwinian theory of evolution. As I interpret that theory it is truly scientific, for the very reason that it does not attempt to explain anything.[1]

For Pearson, curiously, the theory of natural selection can be considered "truly scientific" not because it gives an *explanation* of natural phenomena, but rather because it gives an *accurate description* of them. It is rather debatable, however, whether a descriptive account can be regarded as a scientific theory proper. Are we dealing with any theory at all if the latter does not explain anything? Darwin himself was dubious about that. Sometimes he defines natural selection as a mere "hypothesis,"[2] while at other times he claims:

> In scientific investigations it is permitted to invent any hypothesis, and if it explains various large and independent classes of facts it rises to the rank of a well-grounded theory. The undulations of the ether and even its existence are hypothetical, yet every one now admits the undulatory theory of light. The principle of natural selection may be looked at as a mere hypothesis, but rendered in some degree probable by what we positively know of the variability of organic beings in a state of nature,—by what we positively know of the struggle for existence, and the consequent almost inevitable preservation of favourable variations,—and from the analogical formation of domestic races. Now this hypothesis may be tested,—and this seems to me the only fair and legitimate manner of considering the whole question,—by trying whether it explains several large and independent classes of facts; such as the geological succession of organic beings, their distribution in past and present times, and their mutual affinities and homologies. If the principle of natural selection does explain these and other large bodies of facts, it ought to be received.[3]

What does the word "explain" mean here? And why Darwin feels the necessity to clarify that the principle of natural selection is indeed a well-grounded theory if and only if it "*explains* several large and independent classes of facts"? To understand this, we first need to understand what a valid explanation is from the point of view of modern science.

It is commonly agreed that the birth of modern science coincides with the birth of a new method of scientific inquiry established by Galileian-Newtonian physics. Based on this method, any reference to final-formal causes (such as those that can be found in Aristotle's natural philosophy) must be avoided. Scientific explanations must rely only and solely on the notion of efficient cause.[4] As we will see in this discussion, many contend that Darwin's theory fully meets this requirement. Yet, the more one reads, the more evident it becomes that the principle of natural selection does not remove all traces of purposiveness from the natural world. For one thing, evolution by means of natural selection seems to imply some sort of goal-directed adjustments of life forms, which become increasingly adapted to the environment. Moreover, natural selection itself seems to be tacitly goal-oriented, for the simple reason that it pushes living beings to pursue the very same purpose, reproductive success, always and everywhere. Thus, Darwin's theory appears vitiated, at least partly, by finalistic assumptions—whence his own feeling that the explanatory power of his "hypothesis" might be put in doubt.

A few decades later, Pearson still had some serious doubts about Darwin's theory being a true explanation of evolution; and the same can be said of all those who have sought, one after the other, to dissipate such doubts throughout the twentieth century.[5] Today, little has changed, and debates are more heated than ever: What kind of science is Darwinian biology? What type of scientific explanation does it provide? Can we content ourselves with saying that the correct "description of the nature of Darwin's explanation is 'teleonomy' rather than 'teleology'," or that "the form of the teleological explanation remains, but the terms of the explanations are completely naturalized"?[6] What does "naturalized teleological explanation" mean exactly? And how to interpret the notions of teleology and teleonomy within the framework of a Darwinian approach to natural history?

2.2 Teleonomy

Natural selection can be thought of in various ways. First, we can think of it as a passive effect caused by something else, with the result that the concept is deprived of any explanatory meaning. Second, we can understand natural selection as being an active force that pushes evolution forward—and thus explains the origin of new species—without following any predetermined direction.[7] Third, we can identify natural selection with the pressure exerted on the living by the environment, in which case we should speak, more accurately, of ecological or environmental selection. In the literature, it is not always clear to which interpretation scholars are turning their mind; in particular, the second and the third interpretations of the principle of natural selection are often combined.

Whatever the interpretation, the Darwinian theory about evolution is composed of three basic statements:

1. All living beings are replicators: they are replicas of replicas.
2. Not all living beings achieve the same rates of replication: some reproduce more, others less.
3. Life forms modify their traits and continuously adapt to the environment: some adaptations are better than others.

It is worth emphasizing that (2) concerns *quantitative* data—reproductive success, whereas (3) concerns *qualitative* data—better or worse adaptations. The theory of natural selection does nothing but establish a causal connection between (2) and (3) on the basis of (1). At first glance, it seems reasonable to think that the causal sequence runs from (3) to (2): the better the adaptation to the environment, the higher the reproduction rate. But the Darwinian view of evolution is based on a different hypothesis: the higher the reproduction rate, the better the adaptation. In other words, it is (2) that *explains* (3).

Suppose, in fact, that better adaptations explain higher reproduction rates. If that were the case, it would be impossible to trace all evolutionary episodes back to one universal law, natural selection, and to see all of them as being caused by the same natural force or process. Evolution and the natural world would crumble into many different geographic areas and periods of time, each governed by specific constraints. Ecological systems would determine the course of evolution because only the interactions between life forms and habitats would allow some to reproduce more while condemning others to extinction. The causes of evolution would therefore be multiple and of various natures, bound each time to a certain situation, and there would be no universal cause, no *vera causa* behind the transformation of life forms and the origin of new species. For such a *vera causa* be possible, it must be assumed, on the contrary, that ecological systems are shaped (indeed almost sculpted) by the pressure of one single force, omnipresent in nature. Under this assumption, natural history can be understood as being governed by one single law: natural selection. But the problem is that such a law seems totally unconceivable unless one revives the ancient notion of finalism, or final cause, categorically ruled out by modern science. As Francisco Ayala put it some fifty years ago:

> There are two levels of teleology in organisms. There usually exists a specific and proximate end for every feature of an animal or plant. The existence of the feature is explained in terms of the function or end-state it serves. But there is also an ultimate goal to which all features contribute or have contributed in the past—reproductive success. The ultimate end to which all other functions and ends contribute is increased reproductive efficiency. In this sense the ultimate source of explanation in biology is the principle of natural selection. . . . Reproductive fitness can, then, be said to be the end result or goal of natural selection.[8]

This passage gets right to the heart of the matter. According to Ayala, Darwin's theory entails two kinds of teleological notions: proximate adaptive ends and "the ultimate end to which all other functions and ends contribute." To stress the difference between a teleological explanation of evolutionary processes in terms of proximate adaptive ends and a teleological explanation of evolutionary processes in terms of increased reproductive efficiency, Ayala called the latter a "teleonomic" explanation, neatly distinguished from conventional teleological explanations: "The term teleonomy should be used to explain adaptation in nature as a result of natural selection."[9]

Later on, Ernst Mayr objected that Ayala and others—notably George Gaylord Simpson and Jacques Monod—had misunderstood the concept of teleonomy, originally coined by Colin Pittendrigh. According to Mayr, biological processes can be termed as teleonomic processes only in the sense that these are processes "whose goal-directedness is controlled by a program."[10] Here, the key word is "program," and the key idea is that living beings are comparable with computers: they behave as if they were pursuing a goal, but in the end, they are nothing more than machines governed by programs. "Tentatively program might be defined as *coded or prearranged information that controls a process (or behavior) leading it toward a given end.*"[11]

The quarrel went on for several years. Ayala replied to Mayr, accusing him of providing a circular definition of "teleonomy."[12] Putting aside the many interesting details of this debate, it remains to be seen whether the computer metaphor adopted by Mayr and others really enables biologists to do away with teleological concepts that imply some sort of nonmechanistic tendency toward achieving a certain result. I will return to this matter shortly. For the time being, suffice it to say that I will henceforth give the word "teleonomy" precisely the meaning that Mayr stigmatizes as misleading. This word and its derivatives will express the idea that adaptation in nature can be explained as a result of natural selection. It is (2) that *explains* (3).

2.3 *Vera Causa*

Darwin's theory seems so powerful because it traces the particular (3) back to the universal (2), the contingent to the necessary, the concrete to the abstract, quality to quantity. The main goal of the living is to reproduce; and this goal guides the entire history of Life. In this view, "individuals struggle to increase the representation of their genes in future generations, and that is all"; as a result of this struggle, "organisms become better adapted to their local environments, and that is all."[13] No doubt, for most neo-Darwinian biologists, "that is all," as Stephen Jay Gould says with irony. The problem is whether that is enough.

To decide this, we must take a closer look at the concept that lies at the core of any scientific theory worthy of the name—the concept of cause. It is known that the young Darwin attentively read John Herschel's *A Preliminary Discourse on the Study of Natural Philosophy*. In his autobiography, Darwin ranked this book among those that had influenced him the most.[14] Given the huge body of scholarly literature on every detail of Darwin's life and research, it comes as no surprise that the relationship between him and Herschel has been abundantly discussed.[15] One thing that has caught much attention from scholars is Darwin's reference to Herschel's concept of *vera causa*, or "real cause." In a letter to George Bentham, for instance, Darwin writes:

> In fact the belief in natural selection must at present be grounded entirely on general considerations. (1) on its being a vera causa, from the struggle for existence; & the certain geological fact that species do somehow change (2) from the analogy of change under domestication by man's selection. (3) & chiefly from this view connecting under an intelligible point of view a host of facts.[16]

Here, natural selection is defined as the *vera causa* of evolution. According to Herschel, a *vera causa* is that which explains natural phenomena, in the sense that it produces them.[17] By using the phrase *vera causa* in this letter from 1863, therefore, Darwin is claiming that natural selection explains evolution in the sense that it produces the transformation of species. At the same time, he specifies that the understanding of natural selection as the *vera causa* of evolution follows from its characterization as a "struggle for existence."

When Herschel pored over *The Origin of Species*, he did not find Darwin's hypothesis persuasive. On his copy, sent to him by Darwin, he wrote various comments, pointing out that one cannot boast that the *vera causa* of the origin of species has been discovered unless one clarifies what the *vera causa* of all variations subject to selection is—a mystery that Darwin left unsolved. Herschel, as a historian of biology recently remarked, "rejects the idea that any theory of organic change could possibly be adequate—that the theory could be believed to have produced the observed phenomena—without a description of how the actual process of variation could have produced the history of life."[18]

Based on this and other considerations, some may conclude that Herschel failed to grasp the logic of Darwin's "one long argument," whereas others may contend that Darwin sought in vain to win Herschel's approval, without understanding the latter's insights into the nature of scientific explanations. But the most interesting aspect of the Darwin-Herschel *affaire* probably lies elsewhere and has to do with a crucial lesson that could be drawn from Herschel's observations on the study of natural philosophy:

> When we see a stone whirled round in a sling, describing a circular orbit round the hand, keeping the string stretched, and flying away the moment it breaks, we never hesitate to regard it as retained in its orbit by the tension of the string, that is, by *a force* directed to the centre; for we feel that we do really exert such a force. We have here *the direct perception* of the cause. When, therefore, we see a great body like the moon circulating round the earth and not flying off, we cannot help believing it to be prevented from so doing, not indeed by a material tie, but by that which operates in the other case through the intermedium of the string,—a *force* directed constantly to the centre. It is thus that we are continually acquiring a knowledge of the existence of causes acting under circumstances of such concealment as effectually to prevent their direct discovery.
>
> In general we must observe that motion, wherever produced or changed, invariably points out the existence of *force* as its cause; and thus the forces of nature become known and measured by the motions they produce. Thus, the *force* of magnetism becomes known by the deviation produced by iron in a compass needle, or by a needle leaping up to a magnet held over it, as certainly as by that adhesion to it, when in contact and at rest, which requires force to break the connection; and thus the currents produced in the surface of a

quantity of quicksilver, electrified under a conducting fluid, have pointed out the existence and direction of forces of enormous intensity developed by the electric circuit, of which we should not otherwise have had the least suspicion.[19]

In all likelihood, these lines from *A Preliminary Discourse* did not go unnoticed by Darwin. The key idea put forward here is that the *vera causa* must be viewed as a "force of nature." This force is not visible to the naked eye, it cannot be perceived directly. Nevertheless, the existence of this force can be inferred by the effects that it produces. The paradigmatic example of such an invisible "force of nature" is Newton's gravitational force. We do not perceive the gravitational force, Herschel argues, but only the effects produced by it: "the moon circulating round the earth and not flying off." Along the same lines, Darwin argued some years later that we do not have any direct perception of natural selection—we can only see the effects of it—the transformation of species over time. But the fact remains that natural selection, just like gravity, can be understood as a "force of nature," one that becomes "known and measured," in Herschel's words, by the effects that it produces and hence explains: *vera causa*.

Importantly, in Darwin's view, if natural selection can be thought of as a *vera causa*, this is because it follows "from the struggle for existence," as he says to Bentham. In other words, the characterization of natural selection as a *vera causa* presupposes a dramatic competition among living beings. And this competition, in turn, implies that a tendency toward maximizing reproductive success is present and active in all living beings, which therefore struggle with each other for the purpose of reaching higher rates of reproduction. Absent this pressure toward propagation, we could not speak of "struggle for existence." But what evidence can we provide for the existence of such pressure?

As a matter of fact, when we look at the natural world, all we can see is that reproduction rates vary from one population to another; and all we can do is measure the higher or lower reproduction rates of distinct populations. These measurements, in and of themselves, do not show that there exists pressure toward maximizing reproductive success in all living beings. They just show that some populations reproduce more and others less. Neither these measurements nor any other empirical observation allow us to perceive directly the tendency toward propagation. And yet it is precisely such a push toward reproductive value maximization that makes the "struggle for existence" possible.

Many consequences can be drawn from this remark. The first is that the computer metaphor proposed by Mayr does not convey the exact meaning of Darwin's theory. The "struggle for existence" cannot be understood as a "program." Rather, it is a hypothetical law of nature, from which natural selection follows, based on the assumption that living beings compete with one another continuously because of a push toward reproductive value maximization inherent in all of them—a push that cannot be perceived directly but can nonetheless be extrapolated from the known facts, as Herschel argues. Darwin became convinced that this extrapolation is legitimate after pondering on Thomas Malthus's "principle of population." He thus reached the conclusion that this pressure toward maximizing reproductive success indeed exists. Clearly, this pressure or tendency is not related to, or caused by, any particular features or "programs" of the living; it does not depend on having such and such, species-specific trait or traits. It pertains to all living beings; it is that which defines Life as such, abstracted from the living. In the final analysis, for Darwin, this tendency triggers the "struggle for existence," which then pushes forward the evolution by means of natural selection of ever *newer* life forms with *new*, species-specific traits.

To this, it may be objected that "struggle for existence" is nothing more than a metaphor for Darwin. To be sure, he did not intend to say that living beings literally fight each other in order to reproduce more. Once we realize this, however, nothing changes. The idea of competition for the purpose of ensuring maximum propagation remains paramount to the theory of natural selection, and this idea entails not only that the behavior of all living beings is fundamentally goal-directed but also that such a universal goal-directedness cannot be rendered into any particular "program" of the living, because it pertains to *all* living beings, without distinction. In this respect, Monod is closer to Darwin than Mayr thinks. Monod describes the goal-directedness of life processes in terms of a "fundamental teleonomic project" of all living beings.[20] This means not only that the latter are "objects endowed with a purpose," but also that the universal purpose that all living beings pursue cannot be traced to any particular feature or program, because nowhere in the organism (or the cell, the DNA) can the message "Maximize reproductive success" be singled out. Curiously, Mayr himself points out that "programs are in part or entirely the product of natural selection."[21] But if programs are always the outcome of a selective process, then natural selection itself cannot be a program.

Thus, the tendency or pressure toward maximizing reproductive rates, which is not a program and is not accessible to direct perception, ultimately amounts to a *postulate* of Darwin's theory. Without this postulate, the principle of natural selection does not hold (the "selfish gene" is among the most recent and best-known version of this crucial postulate).[22] And this postulate is hand in glove with another one, concerning the finite volume of the environment in which the "struggle for existence" takes place. It is in fact the resistance offered by a finite environment that transforms the tendency toward maximizing reproductive success into a pressure proper. If reproductive success could be maximized by all living beings without encountering any obstacle, there could never be any "struggle for existence." Rather, we would witness an infinite propagation and multiplication of species that could cohabit all together, however large their number might be. If, on the contrary, species go extinct in the course of natural history, this is because the volume that living beings can occupy each time is limited, and such a finite environmental volume compels them to struggle with each other for the purpose of occupying the largest possible portion of that volume.

The second postulate clarifies further why the tendency toward maximizing reproductive success is the propelling energy that hides behind natural selection: it exerts an active pressure on a finite, limited environment. At the same time, the second postulate casts light on how the environment is to be characterized within the framework of Darwin's theory: the environment is a volume, and nothing more. This volume withstands the pressure toward reproductive value maximization, and that is all. Again, one may wonder whether that is enough. From Darwin's point of view, the quasi-mechanistic interplay between the pressure toward propagation and the limited volume available for propagation *causes* natural selection, which in turn *causes* evolution. This amounts to saying, following Herschel's definitions, that such an interplay should *produce* and hence also *explain* the increased adaptation to the environment that living beings achieve over time, which is one of the effects of natural selection. Darwin does not deny the obvious: those living beings and life forms that reproduce more are the same ones that reveal themselves to be better adapted to the environment. The problem is: Can a quantitative increase produce and explain a qualitative improvement? In other words, can fitness produce and explain—namely, cause—adaptation?

2.4 Adaptive Qualia

Pearson is not the only Darwinian that has doubted the *explanatory* power of Darwin's theory. More recently, Niles Eldredge criticized the "ultra-Darwinian" interpretation of natural selection as "an active determinant" of evolution, contrasting this interpretation with his own understanding of natural selection as "a passive reflection of 'what worked better than what.'"[23] It is not perfectly clear whether Eldredge, who depicts himself as a Darwinian naturalist, would go as far as renouncing the concept of natural selection, but it is nonetheless clear that he denies natural selection any active, causative, or explanatory power. In a more or less distant future, Eldredge seems to suggest, natural selection will suffer the same fate as Newton's gravitation. Today, general relativity says that gravitation is not a primitive force that causes the universal attraction between physical bodies, but rather the effect, or the "passive reflection," of a warping of spacetime. One day, Darwin's and the ultra-Darwinians' conception of natural selection will be overcome by a scientific revolution in the field of biology quite similar to Albert Einstein's revolution in physics.

This may not happen anytime soon, however. As of now, most evolutionary biologists—not only the ultra-Darwinians—agree that natural selection favors those living beings that develop greater Darwinian fitness; that is, the ability to leave viable and fertile progeny. And the problem remains: How can fitness *cause* adaptation? Is there any way to connect a quantitative plus with a qualitative plus? What is the bridge that takes us from "larger" (number of offspring) to "better" (adaptive traits)? It seems no exaggeration to say that the most important debates in twentieth-century evolutionary biology have centered, directly or indirectly, on this problem. In many respects, this problem resembles another one that scientists and philosophers encountered in the early modern period: the problem of secondary qualities. With the rise of the "new science" during the seventeenth century, there appeared to be an unbridgeable divide between the primary qualities of bodies studied by modern physics and all other qualities, such as odors, colors, tastes, and others, that physics does not take into consideration. As Galileo Galilei explained in *The Assayer*:

> To excite in us tastes, odors, and sounds I believe that nothing is required in external bodies except shapes, numbers, and slow or rapid movements. I think that if ears, tongues, and noses were removed, shapes and numbers and motions would remain, but not odors or tastes or sounds. The latter, I believe, are nothing more than names when separated from living beings, just as tickling and titillation are nothing but names in the absence of such things as noses and armpits.[24]

Herschel restated the idea in his study of natural philosophy: The "motion" of physical bodies is the manifestation of the "forces of nature" that modern science can measure and take as an object of knowledge. The implication is that qualities of physical bodies that cannot be measured cannot be considered objects of scientific knowledge; these are not objective, primary qualities but rather subjective, secondary qualities. If we apply the same concept to the study of evolution, it follows that what is to be taken as real and objective is Darwinian fitness (namely, reproductive success) because fitness can be measured, like motion; as for adaptive traits, on the contrary, their qualitative adaptedness cannot be measured unless one finds a way to express it in the mathematical language of fitness.

What is the qualitative difference between the human hand and the human eye? When we look at these adaptive traits (for the sake of simplicity, I speak of the hand and the eye as single entities) from a Darwinian point of view, the qualitative difference between them counts only insofar as it can be quantified in terms of fitness. To draw an analogy, think of a red ball rolling down a green hill: physicists measure the speed of the ball, which is possible because the red color of the ball enables them to distinguish its shape from the green background; but the green and the red colors do not appear in the mathematical formula that describes the movement and speed of the ball. Colors are not primary qualities, so much so that one can even doubt that they are real. Likewise, adaptations are not primary qualities in a Darwinian perspective. And yet, adaptive traits cannot be deemed superfluous or unreal in this case, as colors and odors are for physicists, because the mathematics of fitness is precisely the means by which better or modified adaptations—which appear in the course of evolution—are to be explained.

The resemblance between adaptive traits and perceptions (of odors, colors, and so on) is further confirmed by the fact that living beings adapt better and better to the environment by improving and widening their *perceptual* capabilities, which allow them to conquer larger and larger portions of the environment. For evolutionary biologists, therefore, "the absence of such things as noses and armpits," as Galileo has it, cannot represent the starting point of scientific investigation, for the simple reason that these things are among those that constitute the very object of their investigation. Therein lies the huge, troublesome difference between modern physics and biology. Not only does biology study qualities that are not objective in the physical sense of the word (that is to say, reducible to mechanical processes), but qualities such as adaptive traits cannot be "separated from living beings," as required by the scientific method promoted by Galileian-Newtonian physics. Based on these considerations, the parallelism with secondary qualities seems all the more appropriate. As Gottfried Leibniz remarks when he discusses the irreducibility of perception to mechanical processes:

> We are obliged to admit that *perception* and that which depends on it *cannot be explained mechanically*, that is, by means of shapes and motions. And if we suppose that there were a machine whose structure makes it think, feel, and have perception, we could imagine it increased in size while keeping the same proportions, so that one could enter it as one does with a mill. If we were then to go around inside it, we would see only parts pushing one another, and never anything which would explain a perception.[25]

Leibniz's witty remark applies not only to every explanation of perceptual experiences, or qualia, in terms of mechanical or mechanistic processes but also, *mutatis mutandis*, to every biological conjecture that seeks to explain adaptive traits, or *adaptive qualia*, in terms of mechanical or mechanistic interactions between environmental volumes and reproductive pressure.[26] In either case, a gap remains open between the realm of shapes and motions, on the one side, and the realm of perceptual and ecological interactions between the living and the environment, on the other. Long story short, the scientific method introduced by Galileian-Newtonian mechanics is not conducive to a better understanding of evolutionary processes.

At this point, it is perhaps worth recalling the solution that evolutionary biologists sometimes adopt to solve this difficulty. The solution consists, purely and simply, in weakening the explanatory power of natural selection. Nowadays, several biologists agree that natural selection cannot be seen as the only process that explains evolution. Some draw attention to further causal factors such a genetic drift, mutation, migration, and symbiogenesis; others concentrate on the role played by structural constraints, morphological development, and "exaptations."[27] Neo-Darwinian (or ultra-Darwinian) biologists, on the contrary, hold firm to Darwin's principle and keep "seeking to explain all evolutionary phenomena strictly in terms of natural selection acting on heritable variation within populations."[28] From this monistic point of view, which exalts the *Allmacht* of natural selection, adaptation remains a puzzling problem.

This puzzle lies at the heart of Darwin's theory. For the latter, as already noted, natural selection explains adaptation, not the other way around. This means that natural selection does not follow any predetermined direction, not even the direction that is recommended, as it were, by the environment. If we made the opposite assumption, indeed, we should concede that natural selection *sees* the environment in one way or another, and this concession would contravene Darwin's view, according to which evolution has no direction and natural selection is a blind, unconscious process. One cannot stress enough the importance of this idea in the history of

Darwinism. The idea is that evolutionary theory should forsake teleological arguments and avoid understanding adaptation to the environment as a final cause, or telos, of evolution. From a Darwinian point of view, adaptation is the effect of natural selection, not the final cause, because the true purpose of natural selection is the maximization of reproduction rates. Natural selection is a teleonomic rather than teleological process.

The problem with this approach is that adaptations ultimately become impenetrable. For example, no answer can be found for the simplest question: Adaptations to *what?* To clarify this point, consider an adaptive trait such as greater fat thickness in very cold regions of the world. Does this trait arise so as to protect the living against low temperatures? Or does it arise so as to allow them to store energy and endure a lack of food supplies at higher latitudes? Is this trait an adaptation to climate? Or is it an adaptation to food scarcity in certain periods of the year? As some biologists have pointed out, "to obtain a high degree of confidence that preventing heat loss was the main factor determining fat thickness in a particular species, considerable work would be needed to distinguish this from other potential explanations."[29] And what is worse, an infinite number of alternative conjectures may be made once we start speculating on past environmental conditions in which living beings found themselves struggling for existence; further types of ecological constraints could be taken into account, leading to further ad hoc adaptive explanations.[30] Which is the good one? Greater fat thickness is an adaptation to *what*?

As a matter of fact, nobody can say for sure because natural selection does not explain how living beings adapt to *this or that* feature of their habitat. It simply posits that they always tend to maximize reproductive success, no matter how they achieve this goal. And this "no matter how" turns adaptive traits into adaptive qualia to which, Darwin *docet*, natural selection is blind by definition. Thus, granting for instance that natural selection sees no difference between adaptation A and adaptation B, and granting that both of them ensure survival, the problem arises as to how natural selection will choose between them—namely, produce and explain either one or the other. As this example shows, there appears to be a huge gap between cause and effect here. If the teleonomic process of natural selection is considered a causal explanation for adaptations, this explanation nonetheless fails to explain how natural selection leads, each time, to a specific effect, to a particular *adaptation to*. Natural selection turns out to be a largely (perhaps *too* largely) indeterminate causal power.

In the end, it is we who see ex post the difference between one effect and the other, one adaptation and the other, when we look at the adaptive process from the outside after it has been brought to completion. But from the ex ante perspective of natural selection, this difference cannot be brought into focus: natural selection makes no distinction—and *cannot* make any distinction—between adaptation to cold climates and adaptation to food scarcity. This is why Darwin describes, consistently, natural selection as a "struggle for existence": all that natural selection can be sensitive to is the limited *volume* of the environmental space. This limitation puts living beings in competition with one another. But this does not suffice to explain a *certain* adaptation as being caused by natural selection.

2.5 The Fit of Form and Form

For neo-Darwinian biologists, another thorny issue is the increase in the organization and complexity of life forms that characterizes natural history. Consider, for instance, the evolution of the human eye. If such a complicated organ were the product of natural selection, it would have evolved gradually. That is, the adaptive functionality of the eye—vision—would have increased with the passing of time, one evolutionary step after the other. Yet, leaving aside the question of whether some kind of proto-eye (say, 5 percent of today's eye) might have possessed some kind of proto-functionality (say, 5 percent of present-day visual capability), the problem arises as to how natural selection might have produced the human eye without being sensitive to its adaptive functionality throughout the evolutionary process. Adaptive functionality means adaptive qualia here (namely, adaptive properties of evolutionary traits that cannot be perceived in advance by "the blind watchmaker," or natural selection).[31] Granting that the human eye as we know it is the final, cumulative effect of natural selection, the adaptive functionality of the eye should have increased step by step, following a certain goal-oriented direction without interruption. Yet, if natural selection is blind and insensitive to adaptive ends, how is that possible?

In reality, the very idea of *cumulative* natural selection requires biologists to reintroduce some sort of directionality into evolutionary processes. And this, in turn, compels them to reintroduce teleological concepts into Darwin's teleonomic view of Life. But cumulative selection is nonetheless necessary to explain the increasing complexity of living systems that leads from bacteria to mammals. Perhaps the notion of cumulative selection is of no use when it comes to the rhizomatic evolution of viruses, but this notion is absolutely crucial when it comes to the arborescent evolution of living beings proper.[32] Darwin himself was well aware of this, and for this reason, he describes evolutionary traits as being both "beneficial" (from the angle of reproductive success) and "useful" (from the angle of adaptation). By conflating these two kinds of benefit, Darwin opened the door to a teleological interpretation of his "one long argument."[33]

For contemporary biologists, the need to revive teleological thinking is even more pressing because the notion of selection acting on random genetic variations poses a serious problem from a mathematical point of view. As has been observed, the number of random variations and recombinations that are necessary for evolution by means of natural selection to be possible is so huge that any explanation in terms of trial and error seems blatantly absurd. It becomes less absurd, however, if one posits that cumulative selection is a process that steers toward some distant "target." In this case, only variations that come closer and closer to that "target" will be preserved, one after the other. Repeat this operation millions of times and you obtain cumulative selection. "There is a big difference, then, between cumulative selection (in which each improvement, however slight, is used as a basis for future building), and single-step selection (in which each new 'try' is a fresh one). If evolutionary progress had had to rely on single-step selection, it would never have got anywhere."[34]

The "target" argument has been proposed by Richard Dawkins and illustrates how some of the most ardent advocates of natural selection continue to explain the emergence of complex living systems. They do this by resorting to teleological concepts and arguments—here, the concept of a "target"-oriented process. For some commentators, it is unclear how natural selection can detect the "target" in question.[35] But even more interesting is the fact that Dawkins, too, doubts his teleological argument. As he remarks, indeed, "in real-life evolution there is nothing that corresponds to steering towards some distant genetic target." We should therefore "think of our target as *anything that would improve survival chances*."[36] Based on this premise, once again, it becomes possible to shift from the notion of adaptive target to the notion of teleonomic fitness. For Dawkins, bluntly put, adaptation explains fitness on page 1, but fitness explains adaptation on page 2; and then, adaptation explains fitness on page 3, but fitness explains adaptation on page 4; and so on and so forth.

Flawed reasoning is by no means Dawkins's prerogative. It can also be found in the works of a Darwinian ecologist like George Evelyn Hutchinson, who describes the ecological niche as an *N*-dimensional hypervolume in which every dimension corresponds to a particular environmental condition or resource that allows a given population or species to survive. Within this conceptual framework, the environment in which a species finds itself living is not to be thought of as an ontologically independent and potentially empty habitat. The concept of "vacant niche" makes no sense from Hutchinson's point of view.[37] For him, the ecological niche is just the abstract set of all relevant environmental variables that permit a certain species to exist. The larger the set, the greater the "adaptability" of species; that is, the capability of species to achieve concrete "adaptation."[38]

According to Hutchinson, the idea that the niche is an *N*-dimensional hypervolume is in accord with a well-established principle of ecological research, "the Principle of Competitive Exclusion, which states that in equilibrium communities no two species occupy the same niche."[39] This principle continues to hold when we assume that the niche is an abstract space, or hypervolume, occupied by a certain species with certain needs. But Hutchinson's argument is not without problems. The first thing to note is that, if the ecological niche is not a natural habitat independent from living beings but an abstract hypervolume defined by the species' needs, then every evolutionary step forward marks the birth not only of new species but also of new niches. As Hutchinson points out, moreover, the evolutionary transition from a certain ecological hypervolume to the next one occurs against the backdrop of a wider, all-encompassing "volume," the "biosphere." This is where things get cloudy.

At this point, the question arises as to what causes the interaction between the species, the niche, *and* the biosphere to evolve. For Hutchinson, who is a convinced Darwinian, natural selection explains everything. But there is no real justification for this claim. On the one hand, adaptation requires adaptability, which natural selection increases by broadening step by step the ecological niche;

that is, the hypervolume of variables permitting species to survive. On the other, the biosphere in which this process develops cannot but impose predetermined conditions on the evolution of niches, constantly steering it in certain directions rather than others. Thus, the evolution of niches, that should be explained solely by natural selection, turns out to be dependent on other unknown factors—and much remains unclear about the underlying, critical interaction between the hypervolume of the ecological niche and the surrounding volume of the biosphere in which the niche evolves.

This remark takes us back to Dawkins and some other problems that the neo-Darwinian reading of evolution raises. If evolution is pushed forward by natural selection, which does not select for any specific adaptation to, but only for the purpose of maximizing reproduction and keeping the process of selection running, then why does evolution pause from time to time? In theory, if the neo-Darwinians were right, the evolutionary push would never decrease; yet, the fossil record attests that evolution is a stop-and-go process, and it also seems to suggest that *adaptation to* represents the end of selection in every sense of the word. To deal with this and other difficulties, various solutions have been devised, from evolutionarily stable strategies to optimality models.[40] Such developments, however, do not solve the main problem that neo-Darwinians encounter: How do we keep teleonomic reasoning immune to teleological thinking? Upon reflection, in fact, it appears that neo-Darwinism cannot do without some teleological assumptions, the most important of which may be called the Principle of the Fit of Form and Form.

Over the last decades, much has been written about the Fit of Form and Function. This concept refers to the matching relation between the form of an organ and the function that it exerts more or less aptly within a certain environment. Again, this delicate issue is closely related to the problem of evolutionary adaptation because the function of an organ is normally taken to be the adaptive end, or the telos, which causes evolution to follow a certain direction and organisms to be shaped in some definite fashion.[41] Several biologists and philosophers have deemed it necessary to explain the Fit of Form and Function in teleological terms, making much use of cryptoaxiological notions such as "proper function," "natural norms," and the like.[42] Here, the risk of projecting some anthropormorphic categories onto natural phenomena becomes very high. But leaving aside that risk for a moment, the point worth emphasizing is that numbers of scholars have thus agreed that the teleonomic concept of natural selection does not suffice to explain the miracle of adaptation to.

As already indicated in this discussion, cumulative selection requires the activation of a target-oriented process. And the same can be said of niche evolution because the increasingly complex interactions between the ecological niche and the biosphere cannot be considered the result of a blind competition among living beings. To explain these and other phenomena, biologists must resort to teleological arguments. This means that they must introduce another hypothesis or principle into their theory, the Principle of the Fit of Form and Form, regardless of whether they do it in a conspicuous or covert way. This principle says that evolution by means of natural selection always has a latent direction, which is prescribed by the environment, for it is the environment that *causes* species to evolve in such a way that they conform better and better with the form of the environment in which they find themselves living. Adaptation is the most common definition for this process of growing interaction and adjustment between the two Forms, the living and the environment.

The reason why the Principle of the Fit of Form and Form is never called by its name in the literature is that it necessarily implies that adaptations are, at least partly, the effect of final-formal causes—an implication that contravenes the basic tenets of modern scientific research. To be sure, there would be an easy way to avoid this unfortunate consequence: the obvious solution would be to remove the concept of adaptation from evolutionary theory. For example, if we looked at evolution not from the point of view of certain living beings situated in a certain environment but from the point of view of large ecosystems, there would be no reason to talk about adaptation. Yet this solution would take us beyond Darwin's theory, and probably also beyond all possible reformulations of the theory of natural selection. It is true, in fact, that ecosystems do not *adapt to* anything but rather reorganize while undergoing change. But it is equally true that "ecosystems are not units of selection."[43]

For Darwin, by contrast, the issue was how to explain the origin of adaptations, and he was certainly not the first to address it. In the late eighteenth century, Johann Friedrich Blumenbach had written about the "formative impulse" (*Bildungstrieb*) of organisms, describing it as a teleological drive and claiming that this impulse subjects the organism to the external forces and conditions that continuously readjust the organic shape to make it better and better adapted.[44] A few years later, Georges Cuvier coined the phrase "conditions of existence" and argued that there is "a principle peculiar to natural history. . . . It is that of the *conditions of existence*, commonly termed *final causes*. As no material body can exist, unless it combine all the conditions which render such existence possible, its component parts must be so arranged as to admit of this possibility, not only in itself but in relation also to whatever surrounds it."[45] The notion of conditions of existence (or "conditions of life") then continued to be hotly debated and became critical to many theories about the transmutation of species developed in the first half of the nineteenth century.

In fact, it is against such theories that Darwin takes a stand in the "Historical Sketch" added to the second edition of his magnum opus.[46] As he points out, Étienne Geoffroy Saint-Hilaire "seems to have relied chiefly on the conditions of life, or the '*monde ambiant*,' as the cause of change."[47] Patrick Matthew "attributes much influence to the direct action of the conditions of life."[48] And Robert Chambers, the "anonymous author" of the *Vestiges of Creation*, writes of an "impulse connected with the vital forces, tending, in the course of generations, to modify organic structures in accordance with external circumstances, as food, the nature of the habitat, and the meteoric agencies, these being the 'adaptations' of the natural theologian."[49] For Darwin, neither of these conjectures explain evolutionary adaptations because it is adaptation that explains everything else here. Hence his sarcastic conclusion: "I cannot see that we thus gain any insight how, for instance, a woodpecker has become adapted to its peculiar habits of life."[50]

For his part, Darwin wanted the perspective to be completely reversed: the adaptations of the natural theologian are not an explanans but rather an explanandum. And yet, in spite of Darwin's tremendous efforts to prove the point beyond any reasonable doubt, many of his contemporaries did not find his theory convincing. For them, there remained a gap between theory and reality, between the causal power attributed to natural selection and its alleged effect—adaptations. Karl von Baer, to whom Darwin had paid tribute in the "Historical Sketch," made the strongest objection: in his view, there is no way to disentangle the concept of adaptation from the notion of conditions of life. Natural selection cannot substitute for the Principle of the Fit of Form and Form:

> Bionomic laws are, in von Baer's view, inseparable from the conditions of existence. . . .
>
> Due to the conditions of existence in which they are forced to live and seek their nourishment, animals of diverse origins can assume similar structural characteristics in order to adapt to their environment. . . .
>
> Throughout his examination of Darwin's theory von Baer objected to the notion that chance variations having an internal source independently of any relation to the external environment could ever lead to functionally adapted organisms.[51]

2.6 Negative Selection

Today, as yesterday, it can be doubted that natural selection explains "how, for instance, a woodpecker has become adapted to its peculiar habits of life."[52] As a matter of fact, every time that Darwinian biologists attempt to unravel the mystery of adaptation, teleological explanations are reactivated in one way or another. In this light, one can better understand George Williams's concern: "evolutionary adaptation is a special and onerous concept."[53] For Williams, this meant that the concept of adaptation must be clarified, but not put aside.[54] Indeed, adaptation is what Darwin's theory aims to explain. No surprise, then, that the "adaptationist research programme," from Williams to Dawkins, has grown into the most debated and influential revival of Darwin's own research programme.

In the early twentieth century, the problem of adaptation had already attracted much attention. Sewall Wright and Ronald Fisher, two of the most prominent theorists of the time, did not see things in the same way. Wright, who had coined the concept of "fitness landscape," was inclined to downplay the role played by natural selection and adaptation throughout natural history. As he put it, "the principal evolutionary mechanism in the origin of species must thus be an essentially nonadaptive one." For him, in other words, "evolution depends on a certain balance among its factors": there must be gene mutation, but not at an excessive rate; there must be selection, but not too severe; there must be prevalence of local inbreeding and a certain amount of crossbreeding.[55] Thus, according to Wright's shifting balance theory of evolution, natural selection and adaptation are just two factors of many others that must be taken into account when studying evolution. Fisher's view was closer to Darwin's. For him, adaptation and natural selection are the key concepts. And to understand evolution properly, it is paramount to see adaptation and natural selection as closely related to what Darwin called the "struggle for existence":

> An organism is regarded as adapted to a particular situation, or to the totality of situations which constitute its environment, only in so far as we can imagine an assemblage of slightly different situations, or environments, to which the animal would on the whole be less well adapted; and equally only in so far as we can imagine an assemblage of slightly different organic forms, which would be less well adapted to that environment.[56]

In many respects, this passage illustrates the strategy later adopted by the neo-Darwinian adaptationist programme. In accord with Darwin's original argument, evolution is construed as a competition game between living beings that differ from one another because of slight random (genetic) variations upon which natural selection acts. And the basic idea is as follows: natural selection never says yes; it always says no. This is how neo-Darwinian biologists convince themselves that they can tame the demon of teleological thinking.

In Fisher's view, evolution by means of natural selection does not prompt adaptation directly; rather, some organisms prove to be "less well adapted to that environment" with the passing of time. Therefore, it is not true that natural selection pushes living beings to adapt their organic form teleologically to the form of the environment because natural selection does nothing but cause some of them to distance themselves from the worst adaptations, which prevent other life forms from reaching higher rates of reproduction. Here again, adaptation equals fitness, and teleological explanations are considered reducible to teleonomic calculations. Based on this Mephistophelian premise, Fisher and the neo-Darwinians deem it possible to conclude that evolution does not follow any goal-oriented instructions; and someone like Dawkins can be fooled into thinking that the target argument does not really imply that natural selection works with adaptive ends in view. If anything, natural selection now resembles Walter Benjamin's angel of history: a witness of eternal destruction rather than the designer of a better world:

> His face is turned toward the past. Where we perceive a chain of events, he sees one single catastrophe which keeps piling wreckage and hurls it in front of his feet. The angel would like to stay, awaken the dead, and make whole what has been smashed. But a storm is blowing in from Paradise; it has got caught in his wings with such a violence that the angel can no longer close them. This storm irresistibly propels him into the future to which his back is turned, while the pile of debris before him grows skyward. This storm is what we call progress.[57]

There is no need to stress that the analogy between the angel of history and natural selection must be taken with a pinch of salt, not least because Benjamin's angel of destruction is propelled into the future by an invisible storm that blows in from Paradise, whereas natural selection is considered by Fisher and the neo-Darwinians to be the storm itself. In other words, natural selection does not simply witness evolution, as an angel could do, but rather *causes* living beings to evolve over time; and by saying no to those that it condemns to death, it says yes to the others, the winners. As a matter of fact, it is impossible to do one thing without doing the other at the very same moment. In a competition game, one wins only if someone else loses, *and* vice versa. Viewed in this light, the power of natural selection is diabolical, literally. Not only does it "set something against something," as the Greek word *diaballein* means, but it also misleads those who believe that emphasis on the negative action of selection protects them against the risk of teleological reasoning. In reality, as we learn from Mephistopheles, natural selection shouts its battle cry, "*Ich bin der Geist der stets verneint*," while whispering in a lower voice, "*Ich bin ein Teil von jener Kraft, Die stets das Böse will und stets das Gute schafft*."[58]

To put it in plain English, Fisher's procedure of negative selection (my definition) does indeed entail that no adaptation to the environment is *directly* aimed at by natural selection, but it nonetheless entails that better adaptations to the environment are *indirectly* aimed at, one after the other, by means of exclusion of the "less well adapted" living beings. If natural selection condemns the latter to extinction, this is because natural selection is able to see that the latter fail to conform to the environment, which continues to function as a final-formal cause. From this perspective, therefore, natural selection plays the role of a *vis a fronte*, which is not totally blind but has the capacity to test the waters in one way or another, steering evolution in certain directions and leaving the final word to the Principle of the Fit of Form and Form, which turns the pile of debris accumulated over the centuries—the fossil record—into a perennial monument to evolutionary *progress*.

In addition, it should be noted that the question "Better adapted to *what?*" can be split in two at this point—whence a double-sided perspective that is typical of neo-Darwinism and fraught with consequences. To start with, it is possible to answer that natural selection causes living beings to become better adapted to the environment by causing the "less well adapted" specimens of life to disappear. And on this basis, it can be argued:

> If the environmental states follow at random, selection during one generation cannot prepare the population for the environment of the next. . . . If, however, the environment remains the same for several generations, selection can adapt the population.[59]

Here, it is quite evident that the environment is thought of as a final-formal cause that brings about effects through, or in conjunction with, natural selection. The idea is that natural selection produces adaptation on the condition that the environment remains the same for a sufficient amount of time. If the condition obtains, the environment can be transformed into a stable target by natural selection, which will mold living beings into better-adapted ones.

But this is not the whole story. Indeed, the primary order—or the law that living beings follow throughout evolution by means of negative selection—is the order to maximize reproductive success with a view to achieving this result not only at the present time, but also in a distant future, finding new ways to increase fitness as well as new ways to increase the very capacity to increase fitness. In this case, the answer to the question "Better adapted to *what?*" is no longer the environment but selection itself. And on this basis, Darwinian biologists can argue that natural selection not only causes evolution but also fosters the "evolvability" of living beings, emphasizing this teleonomic property or propensity as one of the most important for evolutionary radiation (or increase in taxonomic diversity):

> The capacity of a lineage to evolve has been termed its evolvability, also called evolutionary adaptability. By evolvability, we mean the capacity to generate heritable, selectable phenotypic variation. . . .
>
> Evolvability may have been generally selected in the course of selection for robust, flexible processes suitable for complex development and physiology and specifically selected in lineages undergoing repeated radiations.[60]

This dual characterization of evolutionary adaptation in terms of adaptation to the environment and adaptation to selection proves, once more, that teleology and teleonomy tend to converge within the framework of neo-Darwinian biology. At the same time, it shows how both teleology and teleonomy contribute to shaping the Darwinian morals of Life.

With regard to teleology, it continues to be critical to all adaptationist versions of evolutionary theory centered on the notion of negative selection. When two alternative evolutionary traits present themselves, it makes no difference whether one opts for A because it is good or discards B because it is bad, for one cannot elude what is worse (B) unless one knows what is better (A). Ultimately, the two choices result in the same decision, which is moral in nature: the *good* choice is the one made by the form of life that respects the aversion of Life for nonadaptive pitfalls.

When it comes to teleonomy, the implications are no less significant. The notion of negative selection entails that there will always be "less well adapted" living beings that are destined to lose the battle for life. But there is more to it than that. Negative selection also entails the never-ending diversification of life forms among which such a battle can be resumed over and over again. Thus, evolution by means of negative selection fuels two tendencies: first, toward improving adaptation to the environment; second, toward diversifying living beings as much as possible.

2.7 Scientific Mythology

Before closing this chapter, it is time to start examining whether Darwinism has not only a scientific but also an extrascientific meaning that makes it particularly attractive and valuable in today's world. In part, the answer has already been given: one of the most striking features of the Darwinian science of Life is the conflation between nomological statements, dealing with facts, and axiological pronouncements, dealing with goals to be valued and pursued by living beings. The axiology of Darwinism stems directly from its teleological and teleonomic assumptions. In fact, the theory of natural selection is vitiated by finalistic assumptions concerning the adaptive ends, or adaptations to, that are achieved throughout evolution. To clear the fog, neo-Darwinian theorists have introduced a new concept, teleonomy, that should explain a number of seemingly teleological processes in terms of quasi-mechanistic processes. But in reality, as highlighted in this discussion, finalistic assumptions are doubled here rather than being eliminated. Teleonomy adds a new telos to evolution without disposing of teleology, even though neo-Darwinian researchers insist that, based on their calculations, 1 plus 1 equals zero instead of 2. The result is that the science of Life morphs into the morals of Life: evolutionary biology claims to be able to know what the living ought to pursue and value, and what they ought not.

The same kind of trompe-l'oeil argument that leads to such an overambitious conclusion can be found in *The Origin of Species*. Some fifteen years before this book was published, Darwin still thought that evolution slows and stops every time organisms become "perfectly adapted" to their conditions of life. As one historian of nineteenth-century biology has stressed, Darwin believed at the time that "perfectly adapted forms are not capable of further improvement."[61] The same historian also underscores that this "assumption of perfect adaptation, which Darwin shared with most of the biologists of his generation, was derived from the belief that nature is a created, harmonious, and purposeful whole."[62] In his 1844 *Essay*, for example, Darwin writes:

> We have every reason to believe that in proportion to the number of generations that a domestic race is kept free from crosses, and to the care employed in continued steady selection with one end in view, and to the care in not placing the variety in conditions unsuited to it; in such proportion does the new race become "true" or subject to little variation. How incomparably "truer" then would a race produced by the above rigid, steady, natural means of selection, excellently trained and perfectly adapted to its conditions, free from stains of blood or crosses, and continued during thousands of years, be compared with one produced by the feeble, capricious, misdirected and ill-adapted selection of man.[63]

Here, what Darwin calls the "trueness" of a race is inversely proportional to its variability and becomes maximal when the race achieves perfect adaptation. In the first edition of his masterpiece, by contrast, Darwin claims that "absolute perfection"—much like the Kingdom of God—is not of this world.[64] Thus, even though he keeps mentioning "the law of the Conditions of Existence," it is now a higher law, natural selection, that explains evolutionary adaptations and the origin of new species. As a result of this change of perspective, what Darwin deems to be "true" is no longer the *form* of life but rather the *force* of Life, which drives living beings to maximize reproductive success and outproduce competing races.

To be sure, when one looks at this process from the angle of its step-by-step outcomes, the impression is that the conditions of life, or final causes, play a role in evolution; and for this reason, Darwin states that "natural selection acts by either now adapting the varying parts of each being to its organic and inorganic conditions of life; or by having adapted them during long-past periods of time."[65] Yet, when one looks at the same process from the angle of the persistent, ineradicable imperfection of all life forms on earth, the impression is that the conditions of life are irrelevant compared to the pursuit of survival and the "struggle for existence," which push evolution forward on a more fundamental level.

As a consequence of this intermingling of teleological and teleonomic arguments, which contradict yet also strengthen one another in the works of Darwin and of his followers, the idea arises that the nomological necessity of natural phenomena is indistinguishable from an axiological imperative that weighs on all living beings: the imperative to achieve survival through adaptation, or adaptation through survival—it makes no difference. Clearly, the suspicion remains that Darwin's theory as well as its various updates are flawed. Some have gone as far as to say that "there isn't any theory of evolution" at the moment.[66] But this leaves open the question of why Darwinism has nonetheless grown into one of the most accredited scientific research programmes of our time, and why the Darwinian view of Life has become influential to such a degree that it is taken for granted by most people today. The most likely reason for this is that the theory of natural selection provides late modern individuals not only with a scientific account of evolution but also with a kind of scientific-mythological narrative that substitutes for old religious beliefs and thus lays the groundwork for a cultural and political reshaping of our societies.

In this regard, the first thing to note is that Darwinism, unlike other scientific paradigms, is often taken to be a final clarification of "the meaning of human existence."[67] This is why neo-Darwinian biologists, just like the early advocates of Darwinism, spend so much time debating with creationists about "the God delusion" or "man's place in nature."[68] Despite appearances, the creationist is the ideal sparring partner for the neo-Darwinian theorist because it is in the high-sounding altercations between them concerning culture, education, and the future of society that the latter finds a way to express the profound message of Darwinism and recall why it matters to all of us. Together, the creationist and the neo-Darwinian evolutionist form a perfect ecosystem, in which Darwin's gospel can be spread all over the globe.[69] The question is: Why do such debates attract so much attention?

The short answer is that in a secular society like ours, in which religious dogmas and principles are no longer on the front page, the need for meaning has not disappeared. What is nature? Where do humans come from? What are the origins of human societies? How should we behave collectively? And what should we value most? To these problems, traditionally addressed by religions and myths, Darwinism gives a new, scientific solution, centered on a new, scientific interpretation of natural history. And we should not be deceived by this novelty. The fact that the Darwinian solution qualifies as scientific does not make it less similar to a religious message or a mythological system of beliefs. Its function is indeed the same one that myths and religion have always had in the past: to allow human beings to find their place in the cosmos. In an age when all religious and mythological illusions have been debunked in the name of enlightenment, this function is taken over by evolutionary biology, which thus turns into the new, secular religion or mythology of our time.

Thanks to this transformation of evolutionary theory into a secular, scientific mythology, the main tenets of Darwinism become almost undisputable, not because they might not be seriously questioned from a scientific or merely logical point of view but rather because biology now serves a function of the utmost importance from a social, cultural, and political point of view: that of providing both scientists and the common people with a creed that enables them to give a meaning to their own existence and to vaguely discern what is good or bad for themselves. The twofold, nomological and axiological, approach to the natural world promoted by Darwinian biology explains why this science—and not, say, physics—has the potential to take a leading role in our societies. It would be blatantly absurd to expect that cosmologists tell us the morals of the Universe. On the contrary, it seems perfectly reasonable to ask Darwinian savants to teach us the morals of Life and divulge the commandments of Mother Nature. The Darwinian science of Life allows them to take on this task.

Over the past century and a half, the morals of Life have thus found expression in the doctrines of social Darwinism, sociobiology, evolutionary psychology, and other trends in the social sciences that have been influenced more or less heavily by evolutionary biology, and that have exerted a heavy impact on the life of Western societies.[70] In the next chapters, I will go over this issue. For the time being, suffice it to say that such an expansion of the Darwinian morals of Life beyond the boundaries of biology brings with it the diffusion of ideas that can be labeled cryptomythological not only because of the role that evolutionary theory plays in today's world but also because of the particular view of nature and human beings that this theory conveys. A case in point is population thinking, which is key to the modern synthesis. According to neo-Darwinism, first, populations struggle for existence; second, competition among populations leads to better adaptations; third, adaptations develop against the background of an ever-increasing adaptability and evolvability of population behaviors, into which the autotelic, self-strengthening power of Life is breathed. Under these assumptions, evolutionary biologists deem it possible to explain a variety of heterogeneous teleological processes (of adaptation) in terms of a single teleonomic process (of selection). But the problem is that, under the same assumptions, some confusion may arise between the concept of individual and that of species.

With regard to individual organisms, they end up losing their ontological integrity. Population thinking entails that only populations can be said to be alive. Living beings are therefore taken en masse and scrutinized collectively. Since natural selection acts solely on the varying traits of competing populations, which all have a life of their own, individuals are not seen as having any identity per se, or any kind of ontological solidity. Rather, each individual organism is considered a transient collection of various adaptive/evolutionary traits/modules that are the ontological features of populations, not of organisms. For neo-Darwinian biologists, in fact, populations are sets of evolutionary traits, not of organisms.[71] As a result, individuals turn into phantomlike entities: on the one hand, they emerge as modular assemblages of evolutionary traits, or adaptive modules; on the other, they all merge into the reproductive substrate, the population, from which new assemblages of new evolutionary traits will originate sooner or later.

With regard to species, they can no longer be understood as natural kinds because population thinking and the related notion that living beings are modular entities contravene any form of essentialism.[72] From this perspective, the very concept of species appears disputable, if not dispensable. But, according to some theorists, it can nonetheless be maintained that if neither species nor individuals possess any life of their own, this is ultimately because species *are* individuals.[73] This conjecture, however bizarre it may seem, is perfectly congruent with the neo-Darwinian approach to biological phenomena, not least because it restates the idea that individual organisms are unimportant when we look at them one at a time. In short, within the framework of neo-Darwinism, either living beings become part of a greater whole, the species-individual, or they disintegrate into smaller adaptive/evolutionary traits/modules that distinguish one population from another and are often identified with so many genes viewed as the units of selection.[74]

There are many similarities between this scientific view and primitive forms of mythic thought: not only the ontological devaluation of individuals but also the belief that collectivities may be thought of as individuals of a higher order—the species-individual on one side, the totem on the other.[75] The most significant similarity, however, lies elsewhere. It is often said that neo-Darwinism naturalizes the human world, and that "Darwin played a major role in moving us to a naturalistic view of human nature."[76] Yet, on close inspection, it is the other way around: Neo-Darwinism humanizes Mother Nature in the first place.[77] Indeed, the self-affirmation and self-celebration of Life throughout evolution by means of natural selection transforms Life into a humanlike divinity that steers evolution toward the fulfillment of teleological and teleonomic goals. Thus, a mysterious—partly comprehensible, partly unpredictable—power ascends to the throne of the natural kingdom. Neo-Darwinian savants, like the revered priests and shamans of ancient civilizations, are charged with the task of decrypting its language and detecting its will. In doing this, they inadvertently revive an archaic mode of thought. Psychiatrists have called it "paleological thinking":

> The paleologician confuses the physical world with the psychological one. Instead of finding a physical explanation of an event, he looks for a personal motivation or intention as the cause of an event. Every act, every event, occurs because it is willed or wanted, either by the person who seeks and explanation or by another person or by something that becomes personified. In other words, causality by logical deduction, often implying concepts involving the physical world, is replaced with causality by psychological explanation, that is, by teleologic causality.[78]

3 Dogmatism, Scientism, and Critical Naturalism

3.1 From Naturalism to Scientism

Naturalism can be defined in two ways.[1] Methodological naturalism says that (1) *natural facts are explained by other natural facts.* Ontological naturalism says that (2) *nothing exists beyond the natural world.* Scientism, which can be considered a distortion of naturalism, goes even further, claiming that (3) *present-day natural science says what Nature is.*[2]

Scientism starts from the premise that the decision about what is real and what is not can only be made by scientists. But the question remains as to who is entitled to be recognized for a scientist. Are there any reliable criteria of demarcation between science and nonscience? Problems like this are well known to philosophers of science, and less so to natural scientists, who are ordinarily more concerned with practical matters—first of all, how to get money for their research programmes—and sometimes accuse others, typically social scientists, of engaging in nonscientific investigations.[3] When this happens, the so-called scientific image of the world risks becoming too rigid because of some preconceived assumptions about the true nature of Nature made by natural scientists. This drift toward scientism can have an impact on the manifest image as well—namely, the everyday representation of human beings and nature.[4] Most of the time, it finds its source of legitimacy in scientific dogmatics.

A case in point is genetics. There is no denying that genes play a critical role in life processes, but when it is assumed that every behavior of the living is controlled or caused by them, the door is wide open to overgeneralizations that lay the ground for two specious versions of scientism: *genetic determinism* and *genetic reductionism.* Based on the alleged axiom that genes explain everything, molecular biologists make a case for more funding, usually overemphasizing the "central dogma of molecular biology," according to which (roughly expressed) the flow of information always goes from the genes to the rest of the organism, and never the other way around. This dogma, to which I will return later, has been called into doubt since the early 1970s.[5] But, for all that, it has not lost ground.[6]

The first thing to note about such a dogmatic belief in the *Allmacht* of genes is that it is far from good for the advancement of science. Quite the contrary, it impedes or slows progress in medical and biological research. Natural science morphs into "ideology."[7] Over the last years, to give just one example, the exceedingly strong emphasis placed on the genetic causes of cancer has diverted attention from other mechanisms that might be at the origins of certain types of tumor, with the result that research teams who wished to test alternative conjectures have been marginalized and denied funds.[8]

The second thing to note is that the gene propaganda confounds the public.[9] To begin with, it becomes difficult for laypeople to assess the actual achievements of research programmes, which are often taxpayer funded. Furthermore, genetic determinism and reductionism fool people into believing that the only way to understand themselves is to look at their genetic makeup.[10] Finally, genetic determinism leads to genetic discrimination; in that respect, the debate that took place in the wake of the publication of *The Bell Curve*, by Richard Herrnstein and Charles Murray, should serve as a warning.[11]

In sum, naturalism does not necessarily mean scientism; scientism does not necessarily mean genetic determinism and reductionism;

but one can hardly deny that the latter are among the most impressive manifestations of scientism today. What goes unnoticed most of the time and deserves some attention is that such forms of scientism are deep-rooted in the modern metaphysics of Life. The intermediate link is neo-Darwinism, which has transformed this metaphysics into scientific dogmatics.

3.2 Historical Limits

As Karl Kerényi points out, the ancient Greeks had two different words for life: *zoe* and *bios*. *Bios* meant life with qualification, such as the *bios politikos* (political life) and the *bios theoretikos* (theoretical life) on which Aristotle expatiated. *Zoe* meant life without qualification, the generic life that is common to all living beings. For the Greeks, therefore, *bios* was the span of a life limited by death, whereas *zoe* was life as opposed to death. The only possible definition of the latter is "not non-life." *Zoe*, as Kerényi notes, "contrasts sharply with *thanatos*. What resounds surely and clearly in *zoe* is 'non-death.' It is something that does not let death approach it. For this reason, the possibility of equating *psyche* with *zoe*, the 'soul' with 'life,' and saying *psyche* for *zoe*, as is done in Homer, was represented in Plato's *Phaedo* as a proof of the immortality of the soul."[12] It is worth noting that the original meaning of the Greek word *zoe* has very little to do with what some philosophers today call "bare life," instantiated by the life of the *Muselmann* in the Nazi concentration camps, who comes closer than anyone else to death.[13] For the Greeks, *zoe* had a festive connotation, pointing to "an amplification of man, in which divine epiphanies are expected and striven for."[14]

The Greek *zoe* has also very little to do with biological Life. As Kerényi explains, "For the present-day student of the phenomenon 'life,' the fact that *zoe* is experienced without limitation is only one of its aspects, not the whole. Here again we cannot speak of a thoroughgoing identity, for, as we have said, *zoe* is the minimum of life with which biology first begins."[15] Thus, according to Kerényi, there is a palpable discontinuity between what we mean by Life today and what the ancient Greeks meant by *zoe* more than two millennia ago. The Greek understanding of *zoe* and our own understanding of Life are divided by "a historical limit."[16] The two words have different connotations, even though they refer, roughly, to the same facts. But which facts are we talking about more precisely? And what is a fact, anyway?

To a first approximation, we may say that a fact is everything that can be explained. If there were no facts at all, there would be nothing to explain. Facts are the sine qua non for the rise of a logos, which is the Greek word for "reasoned discourse" (about something). Viewed in this light, a fact is an explanandum that does not possess any meaning or disclose any logos per se. Rather, a fact always lends itself to various, more or less satisfactory reasoned discourses, or explanations. And a fact regularly brings with it a certain degree of *thaumazein*, of wonder and hesitation, from which knowledge springs.[17] But how many types of explanation can we think of?

As a first approximation, we may say that there as many types of explanation as there are ways to establish a causal link between facts. Aristotle listed four kinds of cause: material, formal, final, and efficient. Immanuel Kant confined himself to discussing the last two, which he termed *nexus finalis* and *nexus effectivus*.[18] For him, only explanations that identify a *nexus effectivus* can be considered scientific in the fullest sense of the word. The reason for this is that the new kind of scientific inquiry established by Galileian-Newtonian science understands natural phenomena in terms of mechanical processes. When studying nature, consequently, early modern science leaves no room for teleological processes or final causes. Early modern science marks the end of Aristotle's natural philosophy. By the same token, it dissipates the confusion created by more archaic, utterly anthropomorphic interpretations of natural processes.

Anthropomorphic thinking reads everything in terms of *nexus finalis* instead of *nexus effectivus*. In the former case, the final cause is a purpose that explains retroactively a certain event, or a sequence of events; the purpose is therefore a *vis a fronte*. In the latter case, things are the other way around. The mechanical effect is determined by a cause that is prior to it; the efficient cause is therefore a *vis a tergo*. Without going into detail, it is all too evident that the purpose is a feature of human activities, for the purpose is, by definition, the object of a preexisting will or intentional attitude. Final causes, thus understood, surely have an explanatory power. Yet it remains to be seen whether they are acceptable from the point of view of modern science. Karen Neander thinks so; she boldly states that "the explanatory power of purposive explanations does not derive from their explicit reference to future effects so much as their implicit reference thereby to past intentional attitudes to those future effects. The explanatory clout comes from an implicit backward reference to prior causes, so these teleological explanations are just a species of ordinary causal explanation after all."[19]

The problem with this argument, quite common in the field of philosophy of biology, is that it confuses the explanatory power of teleological reasoning with its epistemic legitimacy. The overall impression is that teleological reasoning is vitiated by anthropomorphic thinking. As soon as we introduce a telos into our reasoning, we end up implying that some intentional, humanlike force, such as selection, is at work in the natural world and guides evolution toward certain ends, despite all attempts to deny the implication.[20] Neander replies that the problem of anthropomorphism does not subsist because biologists manage to remove any metaphorical reference to humanlike intentions from the concept of selection.

"Natural selection is one type of selection process, which counts literally as such, through the death of a rewarding metaphor."[21] But one may wonder whether the metaphor (of selection) can be put to death without nullifying the explanatory power of natural selection.[22] Moreover, even if it were possible to weaken the metaphor, the fact remains that natural selection can be thought of only by way of analogy, thus projecting some sort of intentional attitude or will onto Mother Nature.

Long ago, John Duns Scotus had already stressed that the concept of natural will, or *voluntas ut natura*, is not to be confused with the concept of the human will, or *voluntas ut voluntas*, because the former is nothing more than an analogy.[23] But the point is that modern science drew its strength precisely from the repeal of this type of analogical reasoning. A historical limit, as Kerényi would say, divides modern science from premodern styles of reasoning.

3.3 *Nexus Finalis*

Today, for the vast majority of scientists, evolution is a fact, that is an explanandum.[24] Darwin theorized that evolution is governed by a law, or a "principle," that he called "natural selection." Here, in line with a naturalistic approach to phenomena, two natural facts are put in relation to each other in such a manner that one explains the other. The natural fact to explain is evolution; that is, the origin of new species and the gradual increase in organization and complexity of life forms and of the entire natural world. The second fact, which should explain the first, is natural selection. New and better adapted species originate by means of variation of traits and selection of those traits that reveal themselves to be more profitable for the living.

But what does this mean exactly? Are the winners of the "struggle for existence" those that simply become better adapted to the environment? If that were the case, it would be mandatory to conclude that natural selection is not a blind power, but rather an intelligent engineer that sees the natural world and operates for the good of the living, designing the latter in such a way that they increasingly fit in with the habitat in which find themselves living. We thereby would introduce a *nexus finalis* into our explanation of evolution. Adaptation would be the telos of all evolutionary processes. No wonder, then, that those who know Darwin's theory best insist that "natural selection favors organisms to the detriment of other organisms, without any further specification. It is we who love to call 'better adapted' what is positively selected, because we like to highlight the positive action of selection."[25]

No doubt, Darwin would have agreed with Edoardo Boncinelli on this point. As he writes at the beginning of his masterpiece, he does not intend to explain selection by means of adaptation; if anything, his aim is to explain adaptation by means of selection. In this regard, Darwin's theory is totally at variance with those advocated by his predecessors, who had exaggerated the importance of the "conditions of life," or final causes, for the transformation and evolution of life forms.[26] More than a century and a half later, Boncinelli restates the concept. And his tone is magniloquent: "this is science at its best."[27] I would rather use the interrogative form: Is this science at its best?

If, by "science at its best," we mean science that rejects all kinds of finalistic explanation, or *nexus finalis*, we may give ourselves the benefit of the doubt. Indeed, if the theory of natural selection really explains natural facts rather than merely describing them, then higher or lower reproduction rates—that is, the living's "fitness"—should explain why and how the living become better and better adapted to the environment. This is what Boncinelli calls the "positive action of selection," and this is what natural selection is thought to be: an action that produces an effect, or a *vera causa*. But this cause is sui generis.

For one thing, higher reproduction rates, occurring later as a result of better adaptation to the environment, should be regarded as the cause of that very adaptation, coming at an earlier point in time. Hence the suspicion that a vicious circle hides behind Darwin's theory. Furthermore, as Boncinelli points out, "it is we who love to call 'better adapted' what is positively selected." In his view, therefore, the phrase "better adapted" could be forsaken, with the paradoxical result that the theory itself would lose, at least partly, its explanandum, for evolutionary adaptation could not be considered a fact.

Finally, and most important, finalism does not disappear when we say that natural selection causes adaptation rather than being caused by it, because reproductive value maximization turns into the fundamental telos of the living, the goal that all living beings pursue. Thus, no matter whether one chooses teleological or teleonomic explanations, Darwinian arguments rely on the crucial assumption that Mother Nature's behavior is characterized by some sort of underlying purposiveness.

3.4 Ontological Fallacy

Darwin's theory of evolution ultimately centers on a teleonomic "force" that belongs to all living beings without being the exclusive property of any of them: Life. One of the main features of this natural force is that it can be measured, as it finds expression in higher or lower reproduction rates. Population genetics and the twentieth-century evolutionary synthesis carried on Darwin's research programme. The result is neo-Darwinism, the version of Darwin's theory that best satisfies a basic requirement of modern science: it makes it possible for biologists to embark on a variety of measurements, starting from the observation of gene propagation and genetic recombination.

This explains why, long before the discovery of the double helix structure of deoxyribonucleic acid (DNA) and other findings in the field of molecular biology, the belief that the gene is key to understanding life processes was already endorsed by many biologists. Then, after Francis Crick and James Watson discovered the DNA structure and proposed the "central dogma of molecular biology," neo-Darwinian biologists welcomed both news and integrated them right away into their revised edition of Darwin's theory.[28] It is worth recalling that Crick, in his famous paper on protein synthesis where the idea of the central dogma was first formulated, used the word "dogma" to emphasize that the unidirectionality of the flow of information from DNA to ribonucleic acid (RNA) and the proteins was nothing more than a theoretical principle, "for which proof is completely lacking."[29] For neo-Darwinians, by contrast, the new dogma was to be taken literally because it represented the longed-for proof that the modern synthesis itself was to be taken as dogma: the gene truly *is* the fundamental unit of natural selection and the locus of Life itself. In a matter of years, the prestige and popularity of scientists like Richard Dawkins grew considerably.[30] During the same period of time, genetic reductionism and determinism received a boost. In hindsight, we can see in this chapter of the history of modern biology a paradigmatic example of dogmatism (in the strictest sense of the word) that leads to scientism.

As already noted, scientism can be considered a distortion of naturalism. Methodological naturalism says that (1) *natural facts are explained by other natural facts*. Ontological naturalism says that (2) *nothing exists beyond the natural world*. The difference between the two versions of naturalism, however little it may seem, is nonetheless of the greatest importance. Methodological naturalism makes sense to those who push scientific research forward, because a naturalistic, disenchanted stance toward reality fosters scientific investigation. Ontological naturalism comes into play and gains ground when scientists are confronted with nonscientists who take a nonempirical stance toward reality and make claims that blatantly contravene science. For instance, ontological naturalism is what methodological naturalism becomes when naturalism is challenged by creationism, because the notion of a Creator takes us beyond Nature and natural facts. To this, naturalists reply that nothing exists beyond Nature. And to substantiate their claim, some of them cannot resist the temptation to go even further and say that (3) *present-day natural science says what Nature is*. Thus, in such circumstance, scientism seems to arise as a reaction or defense mechanism against a dogmatic opponent. In reality, however, for this kind of reaction to be possible, some sort of scientific dogmatics must already be in place.

Dogmatism in science always stems from a "transcendental illusion," as Kant calls it.[31] This illusion comes to the fore every time that it is believed that scientific knowledge traces the boundaries of all that exists. Based on such a transcendental mirage, the question concerning the ultimate nature of Nature is prejudged. Scientific creativity and progress are impeded owing to this fatal mistake. It is one thing to say that *everything in the world is natural*; another to say that *present-day natural science tells us everything about Nature*. Absent a theory of everything, one may argue in line with Kant that nobody can say where Nature begins and ends. Worse still, nobody can say whether Nature is One Thing that can be explained by One Theory. It may well be the case that Nature per se is not such a unified whole, and that natural facts are distributed into regions divided by barriers that dismember the body of what we term *the* Nature (in the singular).[32] The thesis that Nature lends itself to a *reductio ad unum* and an all-encompassing, exhaustive explanation is an a priori, unwarranted assumption that gives way to a transcendental illusion, or an ontological fallacy, as soon as one seeks to prove it beyond a reasonable doubt.

Interestingly, transcendental illusions can also arise within particular fields of research. Here, instead of assuming that the whole of reality has the same ontological ground, it is believed that all the phenomena studied by a certain science are the many and diverse manifestations of One Thing that explains all of them. The history of twentieth-century evolutionary theory attests that biologists are tempted to make this mistake, living the quasi-religious "experience of the Universe as a whole," as Julian Huxley once put it.[33] The current predominant version of Darwinism, which magnifies the role of genes and promotes a reductionist conception of the living, claims that the building blocks of Life (genes) and the underlying cause of evolution (natural selection) have been discovered. But the fact is that this scientific approach is vitiated by a transcendental illusion or an ontological fallacy. Not only are we unable to say that Nature as a whole is One Thing, but the idea that Life is One Thing, or a thing-in-itself, remains highly debatable, being exposed as it is to serious criticism, the most important of which regards the goal-oriented, humanlike behavior of natural selection and the ensuing metaphysical connotation of Life that permeates the neo-Darwinian science of the living world.

For these reasons, a third version of naturalism, which we may call "critical naturalism," states that we should learn to tolerate our present lack of reliable knowledge about the living and evolution without succumbing to dogmatism—that is, without minimizing or concealing our persisting astonishment, *thaumazein*, when we find ourselves placed in front of the fact of evolution. If, by "naturalism," we mean a disenchanted stance toward reality, then there is no other way to be a true naturalist. Scientism stems from the dogmatic belief that natural scientists are entitled to say what Nature is. Creationism stems from the dogmatic belief in intelligent design and the existence of one Creator. Critical naturalism opposes all forms of dogmatism, whether scientific or religious.[34]

II The Morals of Behavior: On Biopolitics

4 Toward a New Biopolitical Regime

Modern political power has two faces: the sovereign and the biopolitical. Hence, one can look at the people who live in modern democratic societies from two points of view. On the one hand, the people are all the citizens who express their general will and exert their sovereign power through political representatives: from this point of view, people form a *collective subjectivity*. On the other hand, the same people also amount to a collection of human beings whose lives are taken care of by the state: from this point of view, they constitute a *collective object*, a population, of which biopolitical power takes control. Thus, in biopolitical democracies such as those that characterize our time, people are both empowered and subjugated. They are empowered insofar as they continue to be nominally recognized as citizens entitled to decide for themselves; but they are at the same time increasingly subjugated, as they are managed as a collection of living beings that must comply ever more rigidly with the panoply of biopolitical measures and provisions that governments adopt day after day. In fact, the impression is that this process of subjugation has gone so far in recent years as to pose an unprecedent challenge to Western democracies. It is nothing less than the key democratic notion of the "autonomous man"—the human being endowed with free will, who is de facto and de jure sovereign over himself—that is called into question as a result of this process of biopolitical subjugation.[1]

To have a better grasp of the situation in which we find ourselves living, the first thing to do is to regard the biopolitical regime as being organized on multiple levels. In this chapter, I will briefly discuss three of these levels—the *economistic background*, the *epidemiological apparatus*, and the *ideological order*—and start to illustrate how the biopolitical regime has changed over the last four or five decades. Then, in chapters 5 and 6, I will broaden the horizon of my analysis and clarify further how the new biopolitical regime fosters the subjugation of people by spreading a new understanding of what it means to be human. What I call the "ideological order of Life" will thus emerge as a major cornerstone of the current biopolitical regime.

Before that, however, it is advisable to highlight the main differences between previous inquiries into biopolitical matters and my own approach. This will also allow readers to see why, despite appearances, there is a close connection between part I of this book, devoted to a critical examination of some crucial assumptions hidden behind the modern science of Life, and part II, devoted to analyzing the impact of such assumptions on the evolution of the biopolitical technologies of power.

4.1 The Unanswered Question

As Thomas Lemke points out in the last pages of his brief history of biopolitical studies, a crucial question has gone unanswered so far. Neither Michel Foucault nor any other critical thinker after him has ever investigated in much detail what biological truths about "life" and "living beings" underpin biopolitical policies and technologies of power. This question, according to Lemke, should stand at the center of a new "analytics of biopolitics":

> Biopolitics requires a systematic knowledge of "life" and of "living beings." Systems of knowledge provide cognitive and normative maps that open up

biopolitical spaces and define both subjects and objects of intervention. They make the reality of life conceivable and calculable in such a way that it can be shaped and transformed. Thus, it is necessary to comprehend the regime of truth (and its selectivity) that constitutes the background of biopolitical practices. One must ask what knowledge of the body and life processes is assumed to be socially relevant and, by contrast, what alternative interpretations are devalued or marginalized. What scientific experts and disciplines have legitimate authority to tell the truth about life, health, or a given population? In what vocabulary are processes of life described, measured, evaluated, and criticized?[2]

The program of Lemke's "analytics of biopolitics"—which also includes an inquiry into the economy of biopolitics (who profits from the regulation and improvement of life processes? who bears the costs? what forms of exploitation and commercialization of life can be observed?), as well as a thorough examination of the new forms of biopolitical subjectivation (how do subjects adopt and modify scientific interpretations of life for their own conduct and conceive of themselves as organisms regulated by genes, as neurobiological machines, as composed bodies?)—should seem obvious to anyone.[3] And yet, even though Foucault prepared the ground and created a method for such research, nobody after him took on the challenge, especially with regard to the first point of Lemke's program concerning the scientific foundations of today's biopolitical practices. Leaving aside the highly controversial reformulation of Foucault's concepts of biopower and biopolitics that can be found in the works of Italian philosophers such as Giorgio Agamben and Antonio Negri, some scholars have attempted to look at the neoliberal economy behind the present-day biopolitical regime, others have explored the unprecedented opportunities for biosubjectivation and biosocialization that the new biotechnologies are believed to create, and yet others have started to analyze the new social imaginaries that characterize our biopolitical times.[4] As a rule, the *Stimmung* of these contributions swings between anger and excitement: anger caused by the overarching effects of biopolitical subjugation that some denounce as intolerable; and excitement nourished by the belief that scientific progress is leading humanity to a new promised land, where humans will become able to improve their living conditions by appropriating the extraordinary—perhaps sometimes overstated—achievements of the life sciences. In many respects, both attitudes seem justified; but they also may be considered baseless in the absence of a deeper understanding of the overall epistemological and axiological framework that undergirds the scientific knowledge of "life" and "living beings," and that now guides the process of biopolitical transformation to which Western societies are subject. An "analytics of biopolitics," in Lemke's sense, is still yet to come.

Some clues about the "regime of truth" that constitutes the theoretical background of current biopolitical practices can be found in those inquiries into biopolitics that are usually neglected by critical theorists. Lemke labels them as "biopolitology."[5] This trend of research, prominent in the Anglophone world, adopts an utterly naturalistic approach to the study of politics and political history. The idea is that the causes and forms of political behavior are to be explained by theories that draw on ethological, genetic, physiological, psychopharmacological, and sociobiological hypotheses. Sociobiology and standard evolutionary theories based on neo-Darwinian assumptions play a paramount role in this context, laying the groundwork for almost all investigations in the field of "biopolitology," whether the object of study is the origin of state and society, the evolution of social behavior, or the practical political problems ("biopolicies") that arise from interventions in human nature and changes to the environment.[6]

In the following, I will not expatiate on such biopolitological investigations, which I do not deem very enlightening. In my view, what deserves close inspection is the syntax of knowledge or the grammar of science, which has gained ground over the years, affecting research in various areas of study, biopolitology included. This grammar finds its source code, so to say, in the life sciences of the twentieth century, from evolutionary theory to epidemiology; but it has also become essential to other branches of science—particularly the social sciences, from psychology to economics—which have exerted a heavy influence on the development of the biopolitical technologies by means of which the human population is governed nowadays.

Neither critical theorists of biopolitics nor biopolitologists have ever focused on this grammar. My aim is to sketch a preliminary analysis of its main features, which may pave the way to further investigation and a more complete exploration of the cultural changes that Western societies are experiencing at the moment. In doing this, I will distance myself from Foucault, whose inquiries into biopolitics are partially outdated, and therefore inadequate for understanding what is going on these days. As a matter of fact, he did not even attempt to examine the type of "systematic knowledge," as Lemke has it, that biopolitical practices require today more than ever before. As a result, Foucault failed to grasp some aspects of contemporary biopolitics that can be brought to light only by taking a broader perspective on the grammar of biopolitical science.

From this point of view, for example, it becomes clear that Foucault was partially wrong when he described the biopolitical regime as a "government of individualization."[7] In reality, this characterization holds for the biopolitical regimes of the past, for the nineteenth century as well as for the early and the middle parts of the twentieth century, but it does not hold for our time. Today, the biopolitical regime follows a diametrically opposite direction, fostering a process of increasing deindividualization that allows the morals of Life to take full control of our behavior. Recent developments in the field of economic thought help us to get a first glimpse of how the new biopolitical regime achieves deindividualization.

4.2 Economistic Background

It is trivial to note that economics is an integral part of modern politics. Ever since Adam Smith, to govern means first and foremost to administer the wealth of a nation, ensuring and possibly increasing its prosperity and well-being. It follows that economic thought lies at the heart of the modern "art of government," as Foucault would say. The principles of economic thought, however, have changed over time. Today, one of them finds expression in the title of Ludwig von Mises's *Human Action: A Treatise on Economics*. Man is action, according to von Mises, and economics—which he considered a special branch of "praxeology"—does nothing but scrutinize actions (i.e., behaviors).[8] The vast majority of contemporary economists would agree on this principle and add that economic behaviors have two basic features: first, they do not imply the existence of fully rational individuals; second, they nonetheless follow some rational, understandable laws, without which there would be no way to predict and govern economic processes.

This amounts to saying that the rationality behind economic behaviors—or the logic that makes it possible to gather economic behaviors into a series of actions, choices, and preferences that lend themselves to theoretical and mathematical interpretations—need not be the rationality of a well-informed, clear-headed individual whose actions, choices, and preferences unfailingly abide by the principle of utilitarian calculation. In the past, this type of ultra-intelligent and hyperaware economic agent, the so-called *homo economicus*, had been the main character of economic theories. Nowadays, on the contrary, most economists give up the idea—all in all, a highly imaginative one—that the world is populated by individuals who act on the grounds of detailed knowledge about and perfect estimation of their own utilities. For one thing, what lies beyond the pleasure principle, and thereby beyond the utility principle, has long since erupted into the field of economic science.[9] Moreover, it is all too obvious that human beings are not ultraintelligent, and that they cannot be thought of as natural-born economists either.[10] For these and other reasons, *homo economicus*, the profit maximizer of yesteryear, has been reshaped into a half-blind player, moved by emotional drivers and affected by cognitive limitations that define "bounded rationality."[11]

On these bases, economic activities are no longer understood as being carried out by clear-sighted individuals. The logic or rationality of economic behavior is now placed elsewhere, in a transindividual space of deindividualized, depersonalized agency. For an economic science to be possible, all we need is a plurality of economic actions, of behaviors indexed as choices or preferences that need not be ascribed to any ideal *homo economicus*. If anything, the object of economics is the *behavioral human being*, understood as a modular assemblage of economic actions. The revealed preferences of people thus constitute the field of economic inquiries that are geared toward tracking down regularity and seriality in economic choices, regardless of the individual's rational capacities. Under this assumption, economists can study the conduct of semirational human beings whose behaviors become relevant to them only when they are taken en masse and inserted into conceptual frameworks that are not sensitive to the singleton's prospects and computations.[12] In sum, the subject of contemporary economics, and therefore the target of the new biopolitical "art of government," are *populational human beings* who are not valued for their personal acumen.

This approach is not without its problems, however, because people who act in the economic arena retain memories of their past behaviors and make memory-based, biased self-evaluations of life satisfaction that can invalidate population-based models of analysis—this being particularly true for democratic contexts in which every taxpayer has a say on public policies and can influence decision-making processes. Hence, there is the need for economists to work out a solution that allows them to read *subjective* self-evaluations of utilities in such a way that policymakers may not be fooled by the public's irrational sentiments and may focus instead on the population's *objective* well-being.

In many cases, according to Daniel Kahneman, Amos Tversky, and others, decisions should be made without taking the public's agreement or disagreement at face value.[13] This does not mean that one should totally neglect the people's subjective (qualitative) self-evaluations of utilities, but rather that such evaluations should be given an in-depth interpretation by turning self-reports into objective (quantitative) and far more reliable evaluations of the population's well-being. Economics is not the only field of research in which the concept of evaluation has become crucial in recent years, as we will see shortly, but this example shows with great clarity what is the exact role that evaluations play in the implementation of the current biopolitical technorationality. Here, the function of evaluation is to consolidate the equation between the behavioral human being and the populational human being through an intermediate link, the *evaluative human being*, who is the subject and at the same time the object of evaluation.

Kahneman's argument speaks for itself. According to him, economists trained in psychology can make accurate measurements of the population's objective well-being, or "objective happiness," based on the subjective self-reports that individuals provide about their real-time satisfaction. Initially, the investigator elicits moment-based self-evaluations about the experienced well-being ("moment utility," in Kahneman's terminology) through appropriate methods (among them, the "affect grid"), leaving out retrospective self-evaluations—namely, memory-based assessments of utilities ("remembered utility")—which are not completely trustworthy from a cognitive point of view. Then, this large collection of individual self-reports is entered into algorithms that compute the degree of experienced well-being in the population at large (after considering further variables, parameters, and residual complications such as the "treadmill effect"). In this manner, it becomes possible to pass from first-person, *subjective* self-evaluations to third-person, *objective* evaluations of well-being or happiness. Leaving aside the details, the important thing to note is that, following this procedure, behaviors are framed in two ways: at the beginning, they appear as the behaviors of individuals who give surface evaluations about their real-time well-being; at the end, they turn into the behaviors of a new kind of populational human being—the target and creature of a new biopolitical technorationality.[14]

4.3 Epidemiological Apparatus

The deindividualizing potentialities of the new biopolitical regime are increased by its various apparatuses, among which epidemiology occupies center stage. The importance of epidemiology in the history of the biopolitical technologies of power from the mid-nineteenth century onward cannot be overrated. Here, I will confine myself to some introductory remarks.

"Apparatus," *dispositif*, is a concept introduced by Foucault. As he explains, an apparatus is an ensemble of discursive and extradiscursive practices that has a major function in the biopolitical era: "that of responding to an *urgent need*."[15] It is well known that Foucault was the first to contrast sovereign power with biopolitical power. In view of the fact that the sovereigns of the past had the right to take life, the sovereign power should be understood, according to Foucault, as the power to "*faire mourir et laisser vivre*" (to make people die and let them live), whereas the biopolitical power should be understood as the power to "*faire vivre et rejeter dans la mort*" (to make people live and disallow life to the point of death).[16] To put it succinctly, the biopolitical power does not fulfill its function by condemning people to death but rather by taking care of human lives by looking after the population's health, hygiene, education, and so forth. For this reason, as Foucault points out, biopower basically means *faire vivre*. Over the centuries, several biopolitical apparatuses have been set up to achieve this end. Among them, epidemiology stands out as the one that "enables one to observe, measure and permanently improve the 'state of health' of the population, in which illness is only a variable that depends on a long list of factors."[17]

Foucault never explored the history of epidemiology in detail. In his works, he paid more attention to the prodromes of public health policies in the eighteenth century and the aberrations of Nazi biopolitics in the twentieth century. However, the history of epidemiology is of the utmost importance for anyone interested in the history of biopolitics, for it sheds light on the evolution of the biopolitical technorationality. Roughly, this history can be divided into two main periods: the classical epidemiology of the nineteenth century and the new epidemiology of the second half of the twentieth century (adumbrated by the Nazi epidemiology of the 1930s).[18] The first period is marked by the epidemiology of mortality, as one may call it, when *faire vivre* merely meant to save people from certain death. The second period is marked by the epidemiology of risk, when *faire vivre* begins to mean making people live a better life. In the latter case, biopower does more than send the message "Do not die." The message is now "Enhance your life." To clarify this point, a brief historical excursus may be of some help.

The classical epidemiology of mortality was born around the mid-nineteenth century, during a series of cholera epidemics in England. John Snow studied the mortality rates in several districts of the city of London and realized that the rates increased in the neighborhoods in which water was being provided by a certain company. Some time later, William Farr, the statistician who created the first national vital statistic system in Great Britain, came to the same conclusion after reviewing data from the Newcastle outbreak: cholera was carried by contaminated water rather than polluted air, as was commonly believed back in those days.[19] Their meticulous investigations may be considered the first example of epidemiological research, which had two goals. From an epistemic point of view, epidemiological research served to establish a causal inference (identifying the water supply as the vehicle of contagion). From a governmental point of view, epidemiological research served to respond to an urgent need (stopping the cholera epidemic by disabling the public water pump situated in Broad Street, as the local authorities did upon Snow's recommendation). Thus, the discipline of epidemiology came into being almost two centuries ago, in the form of a Foucauldian apparatus geared toward deciphering the causes of mortality and lowering mortality rates. *Do not die*.

A step further was taken a century later. Richard Doll and Bradford Hill's studies on the correlation between smoking and lung cancer were instrumental in creating a new paradigm for epidemiological research. Doll and Hill compared the morbidity rates between smokers and nonsmokers, realizing that there is indeed a relationship between smoking cigarettes and becoming ill. However, it was impossible for them to make an etiological inference because the nature of the carcinogen and the pathological mechanism leading from smoking to carcinoma were totally unknown at the time. Still, there was evidence that smoking endangers life. Thus, epidemiological research took another direction. Instead of focusing on a single causal agent, or a one-to-one correspondence between cause and effect (a pathogen transmitted through water had been recognized for the necessary and sufficient cause of the cholera epidemics in London a century earlier), epidemiology began to focus on risk factors, adopting a nondeterministic notion of causation.[20] If one cannot affirm with absolute certainty that smoking cigarettes will cause lung cancer to occur, one can nonetheless establish a correlation between the two phenomena, saying that smoking increases the risk of developing carcinoma of the lungs, and that the greater the number of cigarettes smoked, the higher the risk of getting ill.[21] The difference between the two models of epidemiological research, however minimal it may seem, is paramount and opens the door to the "risk society" of our time.[22]

Once this new type of epidemiological apparatus has been put into action, a myriad of things can fall into the scope of epidemiological monitoring and evaluation. Almost everything can be framed as a potential danger lurking around the corner. As a result, risks appear to be thriving and threatening in every spot of reality, whether physical, mental, or social. Against this all-pervasiveness of risk, which is the ontologically independent entity brought to light by epidemiological statistics, the biopolitical technorationality reacts with precautionary measures that epidemiological research helps to contrive.[23] To govern the risks to which our lives are exposed entails not only fighting certain death by establishing a monocausal inference that may explain a sudden increase in mortality rates, but also fighting future death by combating an invisible enemy that can compromise our biological resources and behavioral capabilities in multiple ways, thereby diminishing our chances of survival in the long term. To govern the risks means, in this sense, to ameliorate our life and its quality, regardless of the meaning that life may have for each of us. A universal imperative thus resounds throughout society: *Enhance your life*. It is never said why people should comply.

The fact that most readers will find this comment misplaced or utterly absurd deserves some attention, for it may be a sign—one among many others—that the populational beings that we have become lately no longer think of themselves as true individuals, autonomous persons, who feel the need to meditate carefully on the meaning and purpose of human existence. Rather, we are likely to regard ourselves as mere living beings that just want to secure biological life and improve their adaptive performance without wondering about futile questions. This attitude is clearly understandable, but it also shows how docile we are. Epidemiological surveys tell us how to protect ourselves against all kinds of risk, and everybody complies right along, pursuing the task with tireless effort. It is hard not to be struck by such an unreflective obedience that seems to conceal some sort of dogmatic creed shared by most people today. Before expanding on that, however, it is worth going over the affinities between the apparatus in the Foucauldian sense and the new apparatus of epidemiological monitoring and evaluation.

For Foucault, as already noted, an apparatus has the task of "responding to an *urgent need*" and consists of both discursive and extradiscursive practices, which jointly give rise to technical protocols of government. Current epidemiological surveys fall within this definition: they respond to an urgent need by measuring risk factors, and they do this by governing the risk. Consider again the original approach introduced by Doll and Hill. They conducted their research by following the evolution of the health conditions of several cohorts of British physicians who used to smoke dozens of cigarettes per day and were never discouraged from putting their lives in danger. The lesson to be drawn from this paradigmatic example is that the epidemiological apparatus evaluates all risks related to people's behavior not only by screening them passively but also by operating on them actively, and sometimes by triggering them, with interventions that modify the distribution of and interactions among things and people, as is typically the case with the evaluation of new drugs and their side effects. In this manner, the epidemiological apparatus governs: it confronts and handles risks to become able to control and tame them.[24] Here, knowledge becomes indistinguishable from a routinized exercise of power over people and environments.

Furthermore, the new epidemiological apparatus governs nowadays in the sense that the results of its constant activities of monitoring and evaluation become essential for the everyday exercise of biopolitical power from a local to a global scale. The evaluations provided by the new epidemiology allow scientists to establish criteria for optimization of people's behaviors; at the same time, they allow biopolitical authorities to justify their own policies and long-term agenda on the basis of such innovative criteria. Optimized behaviors are those that turn out to be more adaptive—that is, less exposed to risks—in the light of epidemiological evaluations. These behaviors are believed to be those that improve the "quality of life," as is often said these days. And the new epidemiological apparatus is precisely the branch of the new biopolitical regime that takes care of our lives by ensuring optimization and betterment of adaptive behavior. For this and other reasons that have become apparent over the last three years, it seems safe to say that epidemiological evaluations set the benchmark for the "governmentality" of our time.[25] Long before the COVID-19 crisis, an epidemiological panopticon had been erected at the center of our societies. The pandemic has just shown how far-reaching the power of epidemiological evaluations has become.

That being said, it is true that not all evaluations made today are epidemiological, strictly speaking, for not all of them focus on risk factors. The university, for instance, is a place where evaluations are becoming inescapable, but such evaluations do not identify the dangers to which a life spent in a higher-education institution may be exposed; here too, nonetheless, evaluations serve to optimize behaviors, which means improving the quality of research and teaching in this case. Likewise, further types of evaluation—such as impact evaluation, process evaluation, and decision-making evaluation—serve to improve the quality of managerial interventions. The same holds, of course, for the evaluation of "objective happiness" devised by Kahneman and Tversky, which should improve the quality of governmental policies. In short, everywhere an evaluation is being made today, it is expected to improve the quality of human behavior. And this means that, whatever the context, evaluation aims at establishing criteria for the optimization of behaviors. The higher the degree of optimization, the higher the quality of research and teaching, of managerial interventions, of governmental policies, and of human life in general. Thus, even though not all evaluations are epidemiological *stricto sensu*, the standard set by the epidemiological apparatus of evaluation remains unchanged and reveals itself to be all-pervading: within the framework of the current biopolitical regime, evaluation is always a tool for governing behaviors with a view to optimizing them, to making them more adaptive.[26]

The latter remark helps us to see why the apparatus in Foucault's sense and the new apparatus of epidemiological evaluation that lies at the heart of the governmentality of our time are partly similar and comparable, and partly not. They are certainly comparable in that both are geared toward *faire vivre* (making people live) through technical protocols of government that respond to urgent needs. However, they are also incomparable, in that the optimization of behavior pursued by the new apparatus of epidemiological evaluation does not abide by the principle of "individualization" that Foucault considers essential to biopolitical power. Today, optimization concerns behaviors taken en masse, as behaviors become relevant to epidemiological research only when they are deindividualized. Population thinking is indeed the distinctive feature of epidemiological research, a feature that it shares with evolutionary biology and the life sciences more generally. But population thinking does not merely imply that the population understood as a collection of individuals becomes the object of knowledge. The implication is much stronger than that: individuals as such are no longer taken into account, for the focus of knowledge is now on traits, biological and behavioral modules that individuals may or may not have in common, but which in any case have a life of their own. Here, it is the population abstracted away from individuals that becomes the object of knowledge and the target of biopolitical optimization.[27] In view of that, it should already be clear why the notion of optimization that is becoming critical nowadays must be neatly distinguished from Foucault's notion of normalization.

As we will see, "normalization" is a keyword in Foucault's research on biopolitics. For him, *laws* coexist with *norms* in modern societies. Laws are issued by public authorities and must be obeyed on pain of punishment, while norms of conduct are discerned by science and inculcated in the population through various means, mainly education. Such norms trace the boundaries of *normality*, which mirror and implement the strict rules that *legality* imposes on human freedom. As a result, Foucault argues, not only legality but also normality contribute to building the iron cage in which human beings end up being trapped in modern times. Normality is, among other things, the very manifestation of mental health. In the modern era, as Foucault emphasizes, to be mad means to be an "abnormal" individual. By contrast, to be "normal" means to be an individual who is recognized as such by society and is entitled to be a *true* individual. In short, normalization entails individualization: the individual, whether "normal" or "abnormal," whether well or ill formed, is the main product of biopolitical normalization according to Foucault.

The problem with this reading of biopolitical modernity is that it applies well to the past but not so well to the present. Today, in fact, the nature of biopolitical power has changed, and the epidemiological management of society endeavors to optimize rather than normalize behaviors on a population scale. Optimization defends society against all risks that might compromise its capacity to live and grow, pushing the population to develop ever-more-adaptive behaviors. Significantly, this can be done without dictating any direction to our lives, without normalizing the individual. Worse still, the whole process requires that the individual leaves the scene.

4.4 Ideological Order

The third level of the new biopolitical regime, the one that makes it more and more efficient as time goes by, is ideology. This concept can be given various definitions. Here, I propose one that I deem suitable for the present situation: ideology is order. First, ideology is order in that it coincides with a certain grammar, or syntax, that determines what can be said and what cannot. Second, ideology is order in the sense that it gives everybody a command that is supposed to come from reality itself. In this sense, ideology is a form of concealed metaphysics.

Take for example the concept of Life. When Foucault examines the role that this concept plays in biology, he describes it as an "epistemological indicator." In his view, the word "life" does not denote any concrete object out there; rather, it functions as a metadiscursive category that serves to delimit and unify the vast field of research named "biology" or the "life sciences": "I would say that the notion of life is not a *scientific concept*; it has been an *epistemological indicator* of which the classifying, delimiting, and other functions had an effect on scientific discussions, and not on what they were talking about."[28] In addition to Life, Foucault mentions three other concepts that took on the role of epistemological indicators in the past: *labor* (in nineteenth-century economics), *language* (in nineteenth-century philology), and *human nature* (in anthropology).[29]

There is, however, a remarkable difference between Life and the other epistemological indicators, a difference to which Foucault does not pay enough attention. Life is in fact a concept that, ever since its emergence, not only has played a role within the field of the life sciences narrowly defined but also has had a considerable impact on other sciences, in particular the social sciences—which is why biologistic theorems have blossomed and gained ground in the fields of psychology, sociology, anthropology, politics, and even economics throughout the late nineteenth and twentieth centuries. And if the concept of Life, unlike the concepts of labor or language, has proved able to colonize other branches of science, this is because Life has never been only an epistemological indicator, as Foucault contends. Life has also played the role of an *axiological indicator*.

In part I of the book, I have illustrated how the biological notion of Life has gained an axiological surplus value that transforms scientific statements into moral assertions, thereby pushing biology beyond itself. From the point of view of evolutionary theorists, indeed, Life is to be understood as a propelling force that drives living beings to optimize adaptive performance and propagate to the greatest possible extent. In this perspective, the message "Enhance your life"—which lies at the heart of the "program" that causes living beings to evolve, as Ernst Mayr put it—can be interpreted as a command or a categorical imperative in which the power of Life, the power that pushes evolution forward, finds expression. The power of Life comes to light only through this order of Life, which all living beings receive and must obey without pause. And this order is always natural and moral at the very same time, as attested by the fact that survival by means of optimization and reproduction is the purpose that all living beings pursue with varying degrees of success: some reveal themselves to be better, or more compliant, than others, who misbehave most of the time. On this basis, a hierarchy among living beings can be established, assessing the degree of compliance with the order of Life.

When it comes to humans, measures to raise the level of compliance can also be introduced if the need arises. This is where the social sciences come to our aid: such measures may affect social, economic, political, educational, or psychological aspects of human life. In the final analysis, every social science can be mobilized for the purpose of reaching a higher degree of compliance with the imperative of biological Life. And this explains, at least partly, the capacity to colonize other branches of science that the concept of biological Life has. The general assumption behind this colonization is always the same: in human societies, the order of Life can be enforced through ad hoc measures devised and prescribed by those social sciences which subscribe to the "program" of Life.

As a rule, biologistic theorems proposed in this or that field of social research (think about the theorems of social Darwinism that became commonplace in Nazi Germany) pave the way to biopolitical measures that are devised and prescribed on the grounds of those theorems (think about provisions for racial hygiene and purity in Nazi Germany). But sometimes it may happen that theoretical formulations are somehow bypassed or reduced to a minimum. In that case, social sciences turn into a direct expression of the biopolitical regime of Life. Theorems give way to practical protocols for interventions that follow a predetermined grammar and are geared toward the biopolitical optimization of human behavior, which science actively pursues.

When this happens, as is often the case nowadays, biopolitical *prescriptions* supplant biologistic *definitions*. Consider the issue of mental health and mental disorders. In principle, a preliminary understanding of mental health should be attained before embarking on a treatment of mental disease, or so people believed in the age of normalization, when health meant approximately the same as normality. In our time, however, neurobiologists, psychiatrists, and psychotherapists are unable to find common ground; they radically disagree about the nature of the human mind, the etiology of mental disease, the interpretation of symptoms, and so on. In this situation, many come to the conclusion that the only source of legitimacy for therapeutic interventions is to be found directly in the order of Life: under the assumption that all living beings are naturally inclined to optimize behavior and achieve a better adaptation to the environment, a number of specialists become convinced that the *prescription* of optimized behaviors provides a proper *definition* of mental health.

For those who cherish this idea, the advantage is that, if they succeed, prescriptions and related treatments seem to provide some sort of a posteriori evidence about the origin and nature of mental disease, regardless of whether such evidence is obtained through medication or therapy.[30] For others, typically psychoanalysts, the problem is that those prescriptions and treatments are grounded on the a priori assumption that optimization and adaptive behaviors are the only goals to achieve in our lifetime.[31] As some psychoanalysts also point out, another problem with such treatments is that therapeutic successes are illusory, or only temporary, in most cases.[32] And yet another problem is that patients are doomed to silence. A cognitive-behavioral therapy for phobia, for example, is meant to expunge the phobic behavior and improve the patient's performance without touching on any further aspects of the latter's personal history and character. Thus, treatments for phobic disorders follow a technical protocol that is deemed valid in all such cases, precisely because not too much importance should be attached to the individual's distinctive personality or unconscious mind (the same can be said of cognitive-behavioral treatments for panic attacks, alcoholism, depression, and so forth). To cut a long story short, these therapeutic protocols apply to the behavioral-populational living being rather than the individual proper. Sometimes a therapist is not even necessary for the implementation of such protocols. A smartphone app may do the job.[33]

No wonder, then, that patients are called "clients" by cognitive-behavioral therapists, for these people purchase a prescription from them rather than starting any in-depth analysis of their lives. Since the mental disorder is thought to be caused by a noxious cognitive-behavioral conditioning that has diminished the living's behavioral capabilities and lowered the "quality of life," healing is achieved through psychagogic techniques that are expected to exert a beneficial counterconditioning. If the prescription and ensuing counterconditioning cause the disorder to disappear, the conclusion is drawn that the latter was indeed caused by some earlier maladaptive conditioning. Hence, from a cognitive-behavioral standpoint, it is prescription that provides a veridical representation of mental health, for it is only by means of prescription that therapists can figure out what the origins and nature of mental disorders are. Here, clearly, to study and treat mental illness means to govern and control people's behavior, to shape their conduct through various techniques (including classical and operant conditioning,[34] reinforcement,[35] exposure,[36] modeling,[37] cognitive revision of maladaptive thoughts,[38] and others).

In view of all that, as we will see later, it seems possible that depression, asthenia, mood disorders, and lower behavioral responses sometimes emerge as a form of intimate protest and resistance against the biopolitical regime and its protocols of optimization, which subject every individual to the deindividualizing order of Life. But it is equally possible that depression, understood as a weakening of behavioral capabilities, is a phenomenon that the biopolitical regime often *hallucinates* to exert its power of counterconditioning—hence the ease with which depression is diagnosed these days. Whatever the case, the logic and practice of optimization harmonize well with the economistic view of society, which remains in the background. If mental health becomes the result of a prescription, this is also because health has become a product, a good to purchase from the relevant resellers, among which we find not only pharmaceutical companies but also therapists who do not restore but rather *manufacture* mental health, boosting people's performance and broadening their behavioral repertoire.

To recapitulate, the economistic background, the epidemiological apparatuses, and the ideological order of Life represent three facets of the present-day biopolitical regime. As already noted, modern societies have always looked at themselves through the lens of economics, but the latter is now integrated into a broader scientific framework that lays the ground for a unified grammar of knowledge that fits in with a new grammar of power.[39] Within this framework, every individual dissolves into a faceless behavioral-populational human being subject to a new kind of biopolitical power and a new kind of scientific scrutiny—the two things at the same time. In that regard, it is worth recalling that Bradford Hill, one of the fathers of the new epidemiology, and Ronald Fischer, one of the fathers of population genetics, had the same mentor, Karl Pearson, one of the founders of the school of biometrics and the author of *The Grammar of Science*, in which it is stated that "the unity of all science consists alone in its method, not in its material."[40] This method of knowledge is key to the biopolitical regime of our time.

5 From Normalization to Optimization

In the early 1970s, Michel Foucault coined the concept of a "normalizing society" for the type of society that has been developing over the last two centuries.[1] Some twenty years later, Gilles Deleuze introduced the concept of a "society of control," which he deemed more suitable for understanding the world of today.[2] Here, I will propose yet another idea, the "society of optimization," and discuss the ongoing transformation of a biopolitical regime à la Foucault into a new ethopolitical regime.

In recent years, the concept of optimization and the word "ethopolitics" have also been used by Nikolas Rose. By "ethopolitics," Rose means all "attempts to shape the conduct of human beings by acting upon their sentiments, beliefs, and values—in short, by acting on ethics."[3] In the following discussion, the word "ethopolitics" is given an altogether different meaning: it denotes a new form of power that rules out the very possibility of any ethics, for it acts directly upon behaviors rather than sentiments or beliefs.

Indeed, the new biopolitical-ethopolitical regime fosters a process of radical depersonalization that disintegrates human beings into behaviors. As a result, in a world like ours, in which people learn to regard themselves as deindividualized entities with no or little moral autonomy, there is limited room for any protest and subjective resistance against power. Even the notion of "resilience," through which Rose hopes to revitalize the idea of resistance,[4] seems to lose meaning, as it presupposes a resilient and responsible person or individual—which is what the new biopolitical-ethopolitical regime is committed to silencing.

5.1 Normalization

In some respects, what I suggest calling a "society of optimization" is akin to what Deleuze calls a "society of control," and the main assumption behind my whole argument is the same as Deleuze's: we no longer live in a disciplinary society. According to Deleuze, Foucault had already perceived and analyzed this transition from an old type of society to a new one:

> Foucault has brilliantly analyzed the ideal project of these environments of enclosure, particularly visible within the factory: to concentrate; to distribute in space; to order in time; to compose a productive force within the dimension of space-time whose effect will be greater than the sum of its component forces. But what Foucault recognized as well was the transience of this model: it succeeded that of the societies of sovereignty, the goal and functions of which were something quite different (to tax rather than to organize production, to rule on death rather than to administer life); the transition took place over time, and Napoleon seemed to effect the large-scale conversion from one society to the other. But in their turn the disciplines underwent a crisis to the benefit of new forces that were gradually instituted and which accelerated after World War II: a disciplinary society was what we already no longer were, what we had ceased to be.[5]

In this passage, Deleuze highlights some key aspects of Foucault's inquiries into biopolitics, but the overall impression is that he reads too much in the latter's works. As a matter of fact, Foucault did *not* perceive and analyze the transition from a disciplinary society to a society of control, nor did he make any conceptual distinction between normalization and control. In *The Will to Knowledge*, to give just one example, he talks about "the new methods of power whose operation is not ensured by right but by technique, not by law but by normalization, not by punishment but by control."[6]

Even though Foucault confounds control with normalization here and elsewhere, not only Deleuze, but also Antonio Negri and Michael Hardt, have claimed that Foucault had already pointed out the difference between the two concepts. Such a misreading is probably due to the fact that also Negri and Hardt mix up things in their analysis of biopolitics and biopower:

> The society of control might thus be characterized by an intensification and generalization of the normalizing apparatuses of disciplinarity that internally animate our common and daily practices, but in contrast to discipline, this control extends well outside the structured sites of social institutions through flexible and fluctuating networks.[7]

Briefly put, from Negri and Hardt's point of view, there is no substantial difference between the society of normalization and the society of control. Ultimately, the transition from one to the other is just a matter of intensification and generalization of the normalizing apparatuses of disciplinarity, that become ever more widespread and expand their power as never before. Here again, control does not contrast with normalization; rather, it simply points to a situation in which normalization is pushed to extremes. And again, it seems that much of what Deleuze had said about the society of control goes unrecognized. In the end, we may ask, is it true or false that "the disciplines underwent a crisis," as Deleuze writes, and a disciplinary society is what we have ceased to be?[8]

To tackle this issue, we need to understand first of all what the concept of normalization stands for exactly. Foucault went over this issue several times in his essays and seminars.[9] For him, to a first approximation, normalization is a process that targets bodies and populations, "disciplining" the former and "regulating" the latter. Normalization is achieved, in fact, through the disciplinary "anatomo-politics" of the bodies and the regulatory "bio-politics" of the populations, which are the two faces of the same normalizing regime of power that shapes society by imposing rigid standards of conduct on people. These standards trace the boundaries of what is to be regarded as normal from the point of view of the whole of society. By means of disciplines and regulations, biopower sets its goals, takes charge of the people's lives and endeavors to make them strictly conform to normality, while pushing the recalcitrant or "abnormal" individuals to the margins. As Foucault emphasizes, biopower is far more intrusive than sovereign power, for the "norm" that normality embodies is far more specific and binding than the law, the command of the sovereign:

> Disciplinary normalization consists first of all in positing a model, an optimal model that is constructed in terms of a certain result, and the operation of disciplinary normalization consists in trying to get people, movements, and actions to conform to this model, the normal being precisely that which can conform to this norm, and the abnormal that which is incapable of conforming to the norm. In other words, it is not the normal and the abnormal that is fundamental and primary in disciplinary normalization, it is the norm. That is, there is an originally prescriptive character of the norm and the determination and the identification of the normal and the abnormal becomes possible in relation to this posited norm. Due to the primacy of the norm in relation to the normal, to the fact that disciplinary normalization goes from the norm to the final division between the normal and the abnormal, I would rather say that what is involved in disciplinary techniques is a normation rather than normalization.[10]

Based on these assumptions, Foucault describes disciplinary power as being highly oppressive. And the obvious implication is that, if normalization (or "normation") reveals itself to be so oppressive, this is because it is often met with opposition from those who are subject to it. Indeed, normalization conforms people to standards of normality that are assumed to be natural, but human beings are not always willing or able to comply with such accepted standards. It follows, according to Foucault, that human beings have the capacity, if not the innate tendency, to oppose the biopolitical regime of normalization. And disciplinary power has to fight against this capacity and tendency on a daily basis.

It seems plausible that Foucault's concept of normalization had been partly inspired by one of his mentors, Georges Canguilhem, who famously argued that medical (typological or statistical) standards of "normality" run the risk of contravening the underlying "normativity" of life; that is, the living's capacity to invent new, original ways to confront illness and cope with environmental challenges.[11] Similarly, from Foucault's point of view, the biopolitical regime of normalization exerts its power on living beings that are spontaneously inclined to find their own way in life, so to say. Hence, Foucault's ill-concealed sympathy for the Kantian ideal of personal "autonomy," his faith in the "impatience for liberty" that fosters critical thinking,[12] and his long-lasting fascination with abnormality and the "lives of infamous men."[13]

The same opposition between the power of normalization and the counterpower of freedom, or autonomy, can be found in Foucault's late investigations into liberalism, but with an important twist: the opposition is now understood as being fostered by liberalism itself. On the one hand, Foucault avers, liberalism is characterized by the "considerable extension of procedures of control, constraint, and coercion which are something like the counterpart and counterweights of different freedoms."[14] On the other hand, as he also points out, we witness "the appearance in this new art of government of mechanisms with the function of producing, breathing life into, and increasing freedom, of introducing additional freedom through additional control and intervention."[15] Thus, within the framework of liberalism, "control is no longer just the counterweight to freedom," but also "becomes its mainspring."[16]

It is not perfectly clear whether Foucault, in his lectures on *The Birth of Biopolitics*, was toying with the idea that human freedom is a historical artifact rather than an innate need of human beings or a Kantian "fact of reason." The impression is that he left both possibilities open in his late writings and teachings, in which he started to investigate the remote history and the very origins of Western subjectivity.

5.2 Optimization

In sum, it is safe to say that no distinction between a society of normalization and a society of control—or a society of optimization—can be found in Foucault's works, except for a short passage to which I will return at the end of this chapter. And yet, one can hardly deny that ours is no longer a *society of the limit* based on disciplinary normalization. If anything, ours has become a *society of the limitless*, as Pierre Legendre and Jean-Claude Milner (among others) have remarked.[17] This means that normalization is no longer on the front page because society now celebrates and promotes the liberation from all forms of disciplinary oppression, capitalizing on the dream of a limitless freedom.

Interestingly, the notion of limitlessness also appears in Deleuze's paper on the society of control, where we read that "in the societies of control one is never finished with anything," and where some emphasis is placed on "the *limitless postponements* of the societies of control."[18] However, to clarify the difference between a society of the limit and a society of the limitless, I will not take my cue from Deleuze's sparse and somewhat sketchy observations on control, nor will I draw upon the arguments made by Legendre and Milner, who are perhaps too inclined to complain about the decline of the West without paying enough attention to the concrete innovations and conspicuous transformations brought about by the new form of power that shapes today's society. Rather, I will look at the current situation from the perspective of optimization, as opposed to normalization. To understand the difference between normalization and optimization, the best thing to do is to start with some mathematical definitions.

The Normal is the average value of a continuous probability distribution. The best-known graphic representation of the Normal is the bell curve, also known as the "Gaussian curve." Here, we find ourselves in the field of statistics. From Adolphe Quetelet through Francis Galton to Karl Pearson, who was among the first to define the Gaussian curve as the Normal curve, the science of normality is concerned with the average man, who is considered the benchmark against which the features and dispositions of all human beings are to be measured and judged.[19] *The Bell Curve*, by Richard Herrnstein and Charles Murray, is one of the latest monuments to this science of normality, which has now lost ground and does not enjoy a good reputation.[20] From this perspective, normality takes on two roles: scientific-descriptive and political-performative. It is possible to merely observe and study normality from a scientific standpoint, but it is also possible to commend normality and transform it into a moral ideal and a goal to be achieved. The very existence of a normalizing society hinges on a cluster of epistemic-political apparatuses that not only scrutinize normality, but also extrapolate a norm, a principle of social management, from it. Thus, the knowledge-power of normality lays the groundwork for a social orthopedics that leads, as Foucault says, to "subjection."[21] Every deviation from the Normal, every "abnormal" individual, looks like a monster from which society must be defended.

The Optimal is an altogether different concept. In mathematical terms, optimization is a method that allows researchers to compute the Optimal (maximum or minimum) value of an objective function under given constraints. This kind of computation is quite ordinary in the fields of information technology and engineering. Mathematical optimization is also key to a discipline called "operational research," or "management science," which studies how to give optimal or near-optimal solutions to complex decision-making problems. Put simply, suppose that you have an objective (such as to reach the nearest door from the place where you are sitting right now), and suppose that there are some physical constraints or obstacles between the start and the finish lines: the Optimal will be the shortest route that takes you to the nearest door. Or, think about the scheduling of flights: given a certain quantity of airplanes, crews, spaces, and time slots provided by airports, optimization techniques enable managers to schedule flights to achieve maximum benefit at minimum cost. As should be evident from these examples, it is not a question of normalization here, for optimization is a method with which a huge variety of heterogeneous problems can be given highly specific solutions, none of which is comparable to another. There are no standard solutions in this case. The Optimal is not the same as the Normal.

Today, not only are optimization techniques used in many contexts for various purposes, but the very idea of optimization is also at the heart of two sciences, economics and biology, which are instrumental in determining how people look at themselves and society. With regard to modern economics, it can be defined as the study of two intertwined optimization problems: how to maximize profit and how to minimize expenditure. Again, the solution to such problems does not involve establishing any norm of conduct, nor does it entail distinguishing between normal and abnormal behaviors. As Lionel Robbins made plain a century ago, *any kind* of behavior can be subject to economic optimization because any human behavior ultimately results from a choice between alternative options, among which one is reckoned more beneficial or less costly than others. This means, for Robbins, that economic science is not charged with the task of judging human behaviors from a moral point of view, nor with the task of ranking human beings according to socially and culturally preestablished criteria. Economic science confines itself to studying economic behaviors. It does not take into consideration the people behind them. For this reason, in a world in which economic optimization becomes essential to the life of society and the principles of economic rationality override all other considerations, there will be less and less room for the view that society must be protected against so-called abnormal individuals or deviant people. In this world, behavior is all that matters, and behaviors are abstracted from people in order to be analyzed in terms of economic benefit or loss. Ethical assessment of individuals and discrimination among them based on conventional standards of social normality, which translate into standards of personal morality, must be left at the door:

> Economists, equally with other human beings, may regard the services of prostitutes as conducive to no "good" in the ultimate ethical sense. But to deny that such services are scarce in the sense in which we use the term, and that there is therefore an economic aspect of hired love, susceptible to treatment in the same categories of general analysis as enable us to explain fluctuations in the price of hired writing, does not seem to be in accordance with the facts.[22]

Clearly, the fact that today prostitutes and porn stars do not incur reprobation but rather curiosity and sometimes maybe even admiration among the public, is already a sign that the social and moral norms of the past as well as the old-fashioned patterns of sexual normalization, on which Foucault placed special emphasis, have lost prestige and are giving way to a thoroughly economistic view of society that is at variance with any social, cultural, or ethical code of personal conduct to which all members of society should conform.

It does not follow, however, that there are no criteria for the assessment of what is "good" or "bad" in the world in which we live nowadays, in the society of optimization. Rather, the point is that such criteria are not to be found in social, cultural, or ethical norms, but only in economics itself and the other branch of science in which the concept of optimization plays a critical role—biology. Here, as already indicated, this concept means two things. First, there is the optimization of genotypic and phenotypic features, which become better and better adapted to a certain function within a certain environment; second, there is the optimization of natural selection, which tends to increase its own capacity to maximize reproductive success throughout the evolutionary process. For most evolutionary biologists, the latter, teleonomic optimization, causes and explains the former, teleological optimization. That is, the principle of survival maximization governs the natural world. In this view, what pushes history forward is the pursuit of survival through better adaptations, which thus become the most relevant criterion for assessing the living's behavior. This is the "metaphysical" principle, as Stephen Jay Gould calls it, that puts us on the path to understanding what is "good" or "bad" in our time, showing at the same time the deep analogies between mainstream biology and economics:

> Why was Darwin so wedded to a principle of maximization that would strike most of us today as both metaphysical and indefensible (ecosystems, after all, can work perfectly well with far fewer species and lower chemical "yield" per spot)? Schweber, I think, has provided the correct answer by stressing Darwin's allegiance to one of the most popular philosophical approaches of his day—the "Benthamite optimization calculus" promoted by Jeremy Bentham, and many other prominent thinkers in several disciplines, as the utilitarian principle in philosophy and political economy, the "greatest good for the greatest number."[23]

5.3 From Individuals to "Dividuals"

The fact that the biopolitical regime of normalization has been superseded by the biopolitical-ethopolitical regime of optimization is attested to by some major changes that have occurred over the past decades: first, the rise of a thoroughly economistic view of society combined with a biologistic interpretation of human existence; second, the decline of ancient, by now outdated forms of social discrimination and the rise of new forms of stigma; and third, the revision of the psychiatric classification system. All these changes give evidence that power is no longer concerned with the normalization of individuals but rather aims for the optimization of deindividualized behaviors.

A thoroughly economistic view of society is based on the idea is that the laws of the economy determine how society functions and develops. All other aspects (history, culture, religion, and so on) as well as the moral values and social norms inherited from the past take second place to economy. The government of society is seen as being, basically, a government of economic processes that pursues the greatest possible growth in wealth and productivity, leaving out of consideration all other issues—whence a sense of loss and spiritual poverty that some feel, as though our world were just the result of a process of sheer annihilation brought about by the brute force of economic materialism. If we live in a neocapitalist or neoliberal society, however, this is not because we have entered a nihilistic age characterized by the twilight of all idols, but rather because people today worship new idols. Neocapitalism is grounded in new dogmas and rituals that exalt the equation between economic and biological optimizations, between the limitless maximization of profit and the limitless maximization of biological resources and behavioral capabilities.

This equation, in turn, is championed by a new ethopolitical regime that does not impose any norms on individuals because it no longer deals with individuals, properly speaking. Instead, it deals directly with behaviors analyzed on a population scale. Thus, the social imperatives and disciplinary apparatuses of yesteryear, which were meant to oblige the individual to conform to the norm, become obsolete. In our time, it is not the individual that matters to biopolitical power. As Deleuze underscores, with the advent of the society of control, individuals are supplanted by "dividuals."[24] And it is only after dividuals have taken the place of individuals that a new form of social order becomes possible.

At that point, what appears to be good has nothing to do with traditional ethics, or with the moral identity of autonomous persons, for "good" now denotes optimization in the economic and biological senses. "Good" is every optimization of the people's biological and behavioral adaptations that ameliorates the economic performance of societies and increases the survival value of populations. Hence, good is no longer the moral quality of a moral decision made by a moral person, but rather the natural-and-moral feature of a certain trait or behavior that can reinforce Life itself, the Life of societies, the Life of populations, *to varying degrees*: "good" means "better than something else."

The neocapitalist management of society abides by this nomological-axiological principle. And, as should be clear now, the difference between the biopolitical regime of normalization and the ethopolitical regime of optimization could not be greater. In the latter case, since "good" does not apply to the individual but to a discrete biological or behavioral module, the same individual can be judged good and bad at the very same moment, depending on the trait or behavior that is brought into focus. On these bases, a new form of social management can develop: power watches over population traits and behaviors, rewarding and punishing the whole of society, or sections of it, according to multiple criteria, but without aiming to conform the individual to any preestablished and fixed social norm. Quetelet's ideal of the average man is forsaken.[25]

As Deleuze explains, the old disciplinary societies fostered two processes: on the one hand, a process of *massification* that led to social homogenization; on the other, a process of *individualization* that led to personal identification.[26] In saying this, Deleuze follows along with Foucault.[27] Disciplinary societies are, for example, the nationalist societies of the nineteenth and early twentieth centuries, in which the ideal of *homo nationalis* took shape.[28] But today, as Deleuze goes on to explain, power no longer "individualizes and masses together." Rather, it promotes a limitless proliferation of choices and preferences (geared toward optimization of profit) through a limitless diversification of biological and behavioral features (geared toward adaptation and survival). Against this backdrop, it becomes extremely difficult to draw a clear-cut line between normal and abnormal individuals, or to agree with Foucault that biopolitical power "categorizes the individual, marks him by his own individuality, attaches him to his own identity."[29] Significantly, even the forms of social stigma change at this point: what is now despicable is not so much a deviation from normality as a decrease in optimality—that is, a lowering of biological potentialities and a narrowing of behavioral repertoires.

Considering this, Foucault's thesis that biopower always results in normalization and individualization must be rejected. Today, biopower does not aim at imposing restrictions on our choices. If anything, it fosters a limitless expansion of human freedom. But the problem is that this freedom does not belong to anyone. This freedom is not the endowment of any individual, of any moral person; in a world of dividuals, individuals are not the cause but rather the effect of their own choices. This means that the very notion of individuality becomes outdated here, for the individual is regarded as an ever-changing combination of biological traits and behavioral patterns that have a life of their own. Thus, the freedom that we enjoy cannot be thought of as a recalcitrant force that is constantly contrasted and domesticated by power and its disciplinary apparatuses, which serve the purpose of making "individuals subjects," as

Foucault argues.[30] In reality, the new kind of freedom that we have conquered lately coexists and grows together with a new kind of power that operates with a view to maximizing the biological and behavioral potential of the dividuals into which we have fragmented. As I argue next, the new biopolitical regime pursues this kind of optimization in two complementary ways: first, it provides tools for the improvement of biological resources; and second, it manipulates the environment to elicit an amelioration of behavioral capabilities. In both cases, freedom is on everyone's lips, and yet freedom is consubstantial with power, which is no longer disciplinary but, ironically, liberating.

5.4 Operational Definitions

The above considerations may seem excessively abstract. I will now give some concrete examples of how optimization goes hand in hand with deindividualization and takes us beyond normalization, starting with an issue to which Foucault returned several times during his lifetime: psychiatric classification.

In the early 1970s, at the time when Foucault was writing the first volume of *The History of Sexuality*, where the "normalizing" effects of the psychoanalytic approach to human sexuality are put into question, American psychiatry, still influenced by psychodynamic theories and categories, was going through a deep crisis. One major cause of concern had been an article published in 1973 by the psychologist David Rosenhan, who had conducted an original experiment the year before. Rosenhan and eight volunteers had pretended to be mentally ill and asked to be hospitalized in twelve asylums. All of them had complained of symptoms of psychic malaise, including hearing voices that whispered the words "empty," "false," "deaf," and others. After hospitalization, they had told the doctors about their past lives without inventing any detail for the purpose of deceiving them. As a result, they were diagnosed as "schizophrenic" and discharged after a couple of weeks, once the doctors had ascertained the remission of symptoms. Rosenhan spread the word about the astonishing results of his experiment. After a while, another research group invited him to repeat the experiment in its own institution. Rosenhan took up the challenge. This time, psychiatrists identified a bunch of "pseudo-patients"—but this time, Rosenhan had not infiltrated anybody into the hospital. Everyone was a "real" patient.

The echo of Rosenhan's study was vast. Many psychiatrists felt insulted by it. The conclusions that he drew from this unsettling experience are as follows:

a) Whenever the ratio of what is known to what needs to be known approaches zero, we tend to invent "knowledge" and assume that we understand more than we actually do. We seem unable to acknowledge that we simply don't know. . . . The facts of the matter are that we have known for a long time that diagnoses are often not useful or reliable, but we have nevertheless continued to use them. We now know that we cannot distinguish sanity from insanity.[31]

b) A psychiatric label has a life and an influence of its own. Once the impression has been formed that the patient is schizophrenic, the expectation is that he will continue to be schizophrenic. When a sufficient amount of time has passed, during which the patient has done nothing bizarre, he is considered to be in remission and available for discharge. But the label endures beyond discharge, with the unconfirmed expectation that he will behave as a schizophrenic again. Such labels, conferred by mental health professionals, are as influential on the patient as they are on his relatives and friends. . . . Eventually, the patient himself accepts the diagnosis, with all of its surplus meanings and expectations, and behaves accordingly.[32]

c) However much we may be personally convinced that we can tell the normal from the abnormal, the evidence is simply not compelling. . . . Thus, notions of normality and abnormality may not be quite as accurate as people believe they are. To raise questions regarding normality and abnormality is in no way to question the fact that some behaviors are deviant or odd. Murder is deviant. So, too, are hallucinations. . . . Psychological suffering exists. But normality and abnormality, sanity and insanity, and the diagnoses that flow from them may be less substantive than many believe them to be.[33]

As these passages from Rosenhan's paper attest, his rebuttal of the diagnostic criteria in use at the time was radical and echoed similar criticisms made by the antipsychiatry movement. In agreement with Thomas Szasz, Ronald Laing, and others, Rosenhan urged psychiatrists to stop using the categories of "normality" and "abnormality." For him, human beings could not be divided into deviant and nondeviant people. Psychopathologists had better confine themselves to being the students of maladaptive behavior, dropping the idea of abnormal individuals. In his own words, when "the origins and stimuli are known and available, discourse is limited to the behavior itself."[34]

Whether or not the Rosenhan experiment had been correctly conducted and reported, the fact remains that its impact on the psychiatric community was huge.[35] The response papers written by some prominent figures of American psychiatry, including the future chief editor of the *Diagnostic and Statistical Manual of Mental Disorders*, could not attenuate concerns about psychiatric misclassification.[36] Thus, in the wake of the Rosenhan experiment, and also under pressure from gay activists that continued to protest vehemently against the classification of homosexuality as a mental disorder, the American Psychiatric Association thought that the time had come for a change.

To begin with, American psychiatrists agreed to disagree: it was impossible for them to share the same "theoretical definitions" of the mentally ill—those definitions that enable professionals to label an individual as abnormal or deviant, but that vary greatly according to the theoretical framework that specialists adopt and deem valid. For this reason, they decided to compile a list of empirical but highly reliable diagnostic criteria based on theory-neutral definitions of mental disorders.[37] The goal was to avoid that the same patients may be diagnosed as schizophrenic by some and nonschizophrenic by others. In this way, American psychiatrists opened the door to the "operational definitions" of mental disorders that were adopted, a few years later, in the third edition of the *Diagnostic and Statistical Manual of Mental Disorders* (DSM-III), edited by Robert Spitzer and published in 1980. According to most historians, the DSM-III marked the end of a tremendous crisis and the birth of a new type of psychiatry.

Two particular events occurred in the early 1970s that served to coalesce these criticisms into widespread discomfort with the DSM-II/ICD-8. These events were the outcry by gay activists against the inclusion of homosexuality as a mental disorder and the publication of an article in *Science* titled "On Being Sane in Insane Places." . . . Despite the demonstration of Rosenhan's methodological limitations, his study brought to light crucial issues concerning classification.

As a result of these criticisms, a revolutionary change occurred in the classification of mental disorders. This change was initiated in the late 1950s, when Joseph Zubin, a prominent psychologist, invited a philosopher of science named Carl Hempel to speak on issues of classification at a meeting of psychiatrists and psychologists. Hempel adopted a logical positivist approach to science. Logical positivism holds that science should be founded on data that are verifiable and factual; data can be used to test scientific theories. Hempel urged mental health professionals to focus on improving the precision with which they defined diagnostic concepts through the use of operational definitions. Operational definitions require that a construct, such as a diagnosis, be defined by the methods in which the construct is determined or measured (i.e., the operational definition of the construct of intelligence is defined by a score on an intelligence test).

> Hempel's ideas exerted substantial influence on the major thinkers of the time who were attempting to wrestle with the issues of classification. . . . In the early 1970s, Robert Spitzer, who was working with Joseph Zubin, was appointed to head the committee to create the third edition of the DSM (i.e., the DSM-III). Spitzer was aware of a paper (Feighner et al., "Diagnostic Criteria for Use in Psychiatric Research," *Archives of General Psychiatry*, n. 26, 1972) by a group of psychiatrists at Washington University in St. Louis, Missouri, who had proposed the use of diagnostic criteria to help clarify the definitions of mental disorders. Diagnostic criteria were specific, behavioral referents that served as decision rules for the characteristics of a disorder that must be present to warrant a diagnosis. The ultimate goal in introducing diagnostic criteria to classification was to enhance diagnostic reliability by increasing the specificity of the category definition. The St. Louis psychiatrists were strongly antipsychoanalytic in their views, and they advocated for a neo-Kraepelinian approach to psychopathology.[38]

This historical account might be enriched with further details concerning the role of psychopharmacological research (and of the pharmaceutical industry) in the revision process, or the emphasis that had already been placed on the operational analysis of psychological terms by previous American psychologists. With regard to pharmacologists and biomedical researchers, they could only welcome symptom-based diagnoses like those promoted by DSM-III because this new kind of diagnostic criteria fostered the idea that a specific drug might be a cure for a specific disease, or symptom cluster. With regard to earlier psychologists, the works of Stanley S. Stevens and B. F. Skinner, the father of radical behaviorism, prove that the operationalist approach was already popular among behaviorists well before Hempel's presentation at the 1959 conference of the American Psychopathological Association. As early as 1945, Skinner had claimed that both operationism and behaviorism pushed mental health professionals to adopt a third-person perspective and discard untestable hypotheses about the first-person experiences, as well as the personal history and the underlying subjectivity of patients. From the vantage point of operationism and behaviorism, "discourse is limited to the behavior itself," as Rosenhan later put it.[39] In other words, for both operationism and behaviorism, the problem is not how to describe the mind or human subjectivity starting from behaviors, but how to explain behaviors starting from other behaviors, thereby leaving no room for the mentalistic vocabulary used by the psychologists, psychoanalysts, and psychiatrists of the past.[40]

To clarify the difference between theoretical and operational definitions, consider the definition of weight. The theoretical definition says that weight is the effect exerted on a body by the gravitational force. On the other hand, the operational definition says that weight is the measure recorded by the scale on which a body has been placed. When it comes to defining human intelligence instead of weight, things do not change. From a theoretical point of view, we can define intelligence as a set of acquired or innate skills that can, then, be further qualified according to one's theory of intelligence. From an operationalist point of view, we can limit ourselves to defining intelligence as a score on a test.[41]

One of the greatest advantages of operational definitions is that they do not need to rest on a validated theory about the essence or nature of the object under scrutiny. Therefore, once psychiatrists have opted for operational rather than theoretical definitions of mental disorders, they can cease quarreling over the true nature of mental illness.[42] If they are radical behaviorists, they can even call into question the very existence of a thing called the "mind." In their view, all one has to do is study maladaptive and harmful behaviors that compromise the everyday performance of human beings without analyzing in depth the latter's subjectivity or, as Rosenhan feared, label them as deviant individuals.

Based on these assumptions, nosological concepts that apply to the individual, such as "neurosis," become useless and can be replaced with new, operationalized constructs, such as "obsessive-compulsive disorder," that apply directly to particular behaviors, or behavioral patterns, abstracted from the individual who hides behind them.[43] As a result of this change in perspective, a new type of clinical approach is likely to gain favor with specialists in mental health—one that no longer classifies people but rather scrutinizes discrete behavioral modules, aiming at their surgical elimination with little or no regard for the subject's "mind" or personal history. As the introduction to the DSM-III overtly states, "A common misconception is that a classification of mental disorder classifies individuals, when actually what are being classified are disorders that individuals have."[44] In line with this programmatic declaration, "the manual framed its diagnoses as discrete, discontinuous entities."[45] The issue for the authors of DSM-III was no longer, "What is behind the symptom?"[46] Rather, it was how to expunge a disorder without taking into account the individual affected by it.

To understand the logic that governs DSM-III and subsequent editions, including DSM-5-TR, think about "phobia" as a concept. The operational definition of a phobic disorder poses no problem: there is a phobia when a strong, irrational fear arises under certain environmental conditions, and a maladaptive behavior (avoidance) follows. Here, both the fear and the behavioral impairment (avoidance) can be observed and measured, but the important thing to note is that "phobia" is no longer defined as a symptom that is related to an underlying "mind," or to an individual with particular characteristics and a personal history that may help to explain his or her suffering. Rather, a phobia is regarded as a behavioral dysfunction with a life of its own, which presents approximately the same characteristics for everybody and which is therefore disentangled from what happened to the patient in the remote past, as though it belonged to an eternal, anonymous present. The same holds when the dysfunction reveals itself to be part of a Humean bundle of behaviors that compose a syndrome with a higher complexity. The individual as a whole, the "mind," does not matter.

On these bases, treatments are devised and prescribed so as to address such units or bundles of maladaptive behaviors without focusing on the overall structure of the patient's "mind." Psychotherapeutic techniques and pharmaceutical compounds for the treatment of phobia merely aim to optimize the patient's behavior. Consider once more the case of the Optimal route, but suppose now that some spiders appear between you and the nearest door. If you suffer from arachnophobia, you will not be able to optimize the route, taking the shortest path to the door. Yet, as soon as you are relieved from the arachnophobic disorder, defined as a temporary "disability," you become able to optimize your behavior again.

One may wonder, of course, whether such therapeutic interventions really address the "mental" cause of mental disorders and thus tackle the root of the problems. But, from an operationalist point of view, this question is misplaced, for the purpose of operational definitions and operationalized diagnostic criteria is precisely to find a way around a question like this, leaving the enigma of the "mental" cause of mental disorder out of sight by bidding farewell to the concept of the "mind." This is the reason why Skinner looked favorably on operationism: not only did he think that the "mind" is a black box, but he was also convinced that we should not care about it because the box is empty, so to speak. Paul Meehl, by contrast, immediately sounded a note of caution and warned that DSM-III could fool specialists into believing that the definitions listed therein tell the truth about mental disorders. Thirty-eight years after his article on DSM-III, and more than ten years after the release of DSM-5, his warning has lost none of its topicality:

> The extreme (simplistic, "vulgar operationist") form of this view is that the very *meaning* of the concepts is contained, exhaustively and explicitly, in the "operational definitions" provided by DSM. It would be hard to find one single logician or historian of science today (or for that matter, since around 1935!) who would countenance the conception of scientific method enshrined in this view. I find

> it puzzling that physicians, or for that matter, psychologists, unless they are of the most dogmatic behaviorist kind, should adopt this position when neither the history of organic medicine, nor of genetics (I don't mean here merely behavior genetics), nor of traditional trait theory in academic psychology, nor of classical psychometrics gives any support to it. . . .
>
> Nothing but dogmatism on the one hand, or confusion on the other, is produced by pretending to give operational definitions in which the disease entity is literally identified with the list of signs and symptoms. Such an operational definition is a fake.[47]

Today, contrary to Meehl's wishes, dogmatism and confusion continue to surround the operational definitions listed in DSM-III and subsequent editions. As has been pointed out lately, the belief that the DSM constructs "represent the truth" continues to be endorsed—sometimes explicitly, but more often implicitly—by many specialists.[48] As a result, a deinvidualizing approach to mental disorder continues to be widely adopted. The situation has not changed with the release of DSM-5.[49]

5.5 From Freedom to Depression

As already noted, the operationalist approach to mental disease, unlike the disciplinary approach to human conduct, does not rely on the idea that human freedom must be regimented in one way or another. Quite the contrary—among mental health professionals, and more generally within the framework of the current ethopolitical management of society, the common assumption is that the spectrum of behavioral choices must be broadened as much as possible, and the quality of behavioral performance must be improved at all costs, with the goal of making human beings free and powerful as never before.

The current use and abuse of a drug like Viagra offer a simplified model for understanding how the operationalist approach has changed the everyday view of disease, stripping the traditional distinction between sanity and insanity of all meaning. On the one hand, this drug is meant to be a treatment for erectile dysfunction and should be prescribed only for this purpose, regardless of whether the dysfunction is caused by physical (aging) or psychological (anxiety) factors, because the operational definition of the dysfunction does not require physicians to discover its ultimate cause and nature. On the other hand, it is well known that the target market of Viagra is much larger than that, consisting mostly of male adults who do not suffer from any dysfunction but take Viagra for recreational purposes, often mixing it with illicit drugs, to render their sexual performance more durable and satisfactory.[50] Whatever the case, Viagra helps to enhance the behavioral capacities of human beings. Thus, it broadens human "freedom" rather than imposing limits on it. And the question is: Can Viagra be considered a means to normalize sexual performance? In reality, it seems safe to say that this drug makes it difficult to understand where normality begins and ends.

Wasn't it *normal* for sixty-year-old men to experience lower sexual performance until not so long ago? Curiously, since Viagra and similar drugs are expected to expand the range of behavioral possibilities at one's disposal no matter what the initial conditions are, it may happen nowadays that men of all ages feel like they are flawed and disabled, even when they do not suffer from any actual dysfunction. Granting that the operational definition of erectile dysfunction does not trace it to any physical or psychological cause, but rather identifies it as a generic inability to attain or maintain an erection adequate for pleasing sexual activity, these drugs are likely to create a minus while promising a plus, fooling many into believing that sexual behavior and pleasure can be enhanced without limits, and the absence of such a constant enhancement may be a sign of dysfunction: "Men who use Erectile Dysfunction Medications for recreational purposes may be at increased risk of becoming psychologically dependent, which in turn could lead to psychogenic-based Erectile Dysfunction symptoms."[51] As this trivial example shows, the logic of optimization takes us beyond the logic of normalization by canceling the limits of normality and blurring the line between sanity and insanity.

Another telling example of how normalization gives way to optimization is offered by the widespread use among American students of Adderall and similar pharmaceutical treatments for attention deficit hyperactivity disorder (ADHD). In general, the problem posed by such drugs is that they seem to blur not only the line between sanity and insanity but also the line between normalization and optimization. At first glance, one cannot say whether these "smart pills" are meant to normalize or optimize cognitive abilities. As a medical anthropologist has observed, the answer seems to vary depending on the circumstances: "For some, the Adderall experience produces new possibilities to access hidden academic and social potentials"—optimization. "For others, it is evidence that their brains are in fact diseased and they need medication to cope with the demands of college life"—normalization.[52] That said, in the age of the University of Excellence, as Bill Readings defines the higher-education institutions of our time,[53] when the parameters of academic success become higher and higher, everybody's brain runs the risk of being diagnosed as diseased at some point in life. For this reason, it makes little sense to speak of normalization these days, for there are no rigid patterns of normality to which people should conform their behavior once and for all. On the contrary, it is as though the very criteria of normality were being optimized day after day. Against this background, it comes as no surprise that the Adderall epidemic has rapidly spread from college to elementary school, as a correspondent at *The New York Times* reported some years ago:

> When Dr. Michael Anderson hears about his low-income patients struggling in elementary school, he usually gives them a taste of some powerful medicine:

Adderall. The pills boost focus and impulse control in children with attention deficit hyperactivity disorder. Although ADHD is the diagnosis Dr. Anderson makes, he calls the disorder "made up" and "an excuse" to prescribe the pills to treat what he considers the children's true ill—poor academic performance in inadequate schools.[54]

As this episode attests, the imperative of optimization weighs on everyone, and even children are constantly incited by parents to enhance their capacities and bet on their human capital, as Gary Becker would put it.[55] In all corners of society, human beings are urged to improve their behavioral performance to the greatest possible extent—whence a growing anxiety, often combined with a sense of incapacity and sadness that may end in depression.

Depression may be regarded, in fact, as the reverse side of optimization. As already noted in this discussion, depression is the mental disorder that is most likely to be diagnosed when someone feels unable to comply with the rules of the ethopolitical regime and the order of Life that prescribes the never-ending, limitless optimization of human behavior. Not so long ago, the World Health Organization (WHO) announced that depression has slowly grown into a worldwide illness. Surprisingly, however, the definition of depression given by the WHO is far from satisfactory. Depression is described, very superficially, as a condition that prevents people from having a productive life. For this reason, it is considered "a common illness." Two weeks of sadness and inactivity are enough to make anybody depressed:

> Depression is a common illness worldwide, with an estimated 3.8% of the population affected, including 5.0% among adults and 5.7% among adults older than 60 years. . . . During a depressive episode, the person experiences depressed mood (feeling sad, irritable, empty) or a loss of pleasure or interest in activities, for most of the day, nearly every day, for at least two weeks. Several other symptoms are also present, which may include poor concentration, feelings of excessive guilt or low self-worth, hopelessness about the future, thoughts

about dying or suicide, disrupted sleep, changes in appetite or weight, and feeling especially tired or low in energy.[56]

The "two weeks" clause that is essential to making depression such "a common illness worldwide," or "a common mental disorder," comes directly from DSM-III, in which all clear-cut distinctions between melancholia (major depressive disorder) and reactive (neurotic) depression had been abolished.[57] The consequence of this highly controversial decision by the editors of DSM-III is that depression has been overdiagnosed over the last four decades, and drugs for the treatment of depressive disorders have been overprescribed:[58]

> The DSM's diagnostic criteria for Major Depressive Disorder [MDD] abandoned the separation of melancholic and reactive depressions that had persisted for centuries. Instead, it combined them into a single, extraordinarily heterogeneous diagnosis. Someone who had been severely depressed for years, could not leave her bed, and had continuous thoughts of worthlessness had MDD, as did an adolescent who felt depressed and unable to feel pleasure, had trouble sleeping, and lost his appetite and concentration after his girlfriend broke up with him two weeks before.[59]

Based on this overgeneralized definition of depression, the WHO has reached the conclusion that millions of people around the world are affected by this mental disorder, which is considered nowadays the leading cause of so-called disability for both males and females.[60] And it is certainly worth emphasizing that the concept of disability is characterized by the same vagueness, and therefore lends itself to the same abusive overgeneralization, as the concept of depression.[61] As the WHO itself overtly admitted on its website some time ago:

> Disabilities is an umbrella term, covering impairments, activity limitations, and participation restrictions. An impairment is a problem in body function or structure; an activity limitation is a difficulty encountered by an individual in

executing a task or action; while a participation restriction is a problem experienced by an individual in involvement in life situations.[62]

More recently, the WHO has added that "disability is part of being human. Almost everyone will temporarily or permanently experience disability at some point in their life."[63] Thus, both depression and disability end up hinting at a widespread condition that defines the existence of the human population under the new ethopolitical regime: disability is understood as the opposite of a well-functioning organism that has to deal with a variety of environmental conditions; depression is understood as the opposite of mental well-being; that is, the opposite of the ability to take initiative, interact with the environment, and engage in activities that will ensure survival.

Viewed in this light, disability and depression seem more than just two impairments among many others, for they turn into the eternal antagonists of the ethopolitical regime. They become a symbol of all that optimization counters, of all that the society of optimization rejects, of all that is at odds with the new biopolitical regime. But they also foreshadow our fate: in a world where there are no limits to the daily enhancement of human abilities, and where every record can be broken, everybody is doomed to suffer from depressive and disabling disorders sooner or later.

5.6 Foucault and Ethopolitics

To recapitulate, Foucault's categories are not appropriate for understanding the transition from a society of normalization to a society of optimization. New concepts and distinctions are needed, such as those listed here.

First, in the society of normalization, power adopts a disciplinary approach to individuals; in the society of optimization, it adopts an operationalist-behavioristic approach to dividuals.

Second, in the society of normalization, the disciplinary apparatuses restrict and regiment human freedom; in the society of optimization, the ethopolitical apparatuses exalt human freedom in the sense that they broaden the range of behavioral options, thus multiplying the choices at one's disposal. Ethopolitics marks the triumph of capitalism.

Third, in the society of normalization, human beings are subjugated through various techniques of individualization, which force them to conform to some preestablished social norms and relatively rigid standards of conduct. Here, resistance against power takes the shape of illegalism, madness, and other forms of individual "abnormality." In the society of optimization, human beings are depersonalized, in that behaviors are abstracted from individuals. Ethopolitical optimization acts directly on behavioral modules, leaving aside the individual, to make them more adaptive and effective. Accordingly, resistance against power takes new directions, such as idleness (*désoeuvrement*), but also antiutilitarian and self-harmful conduct (as exemplified by rave parties, sadomasochism, extreme sports, and gambling).

Foucault was no stranger to these forms of resistance, and we may even wonder whether his late investigations into cynicism and ancient ethics were an instinctive reaction against the new ethopolitical regime. As already noted, he began to study the mechanisms of biopower in the early 1970s, right at the moment when biopolitics was about to morph into ethopolitics. For this reason, he cannot be reproached for not having perceived that Western society was entering a process of massive transformation at the time. There is, moreover, a short passage in which he seems to realize with some degree of clarity that the situation was rapidly changing. This passage adumbrates the rise of a society of optimization, as opposed to a society of normalization. It is taken from the last cycle of lectures that Foucault devoted to the analysis of biopolitics:

> You can see that what appears on the horizon of this kind of analysis is not at all the ideal or project of an exhaustively disciplinary society in which the legal network hemming in individuals is taken over and extended internally by, let's say, normative mechanisms. Nor is it a society in which a mechanism of general normalization and the exclusion of those who cannot be normalized is needed. On the horizon of this analysis we see instead the image, idea, or theme-program of a society in which there is an optimization of systems of difference, in which the field is left open to fluctuating processes, in which minority individuals and practices are tolerated, in which action is brought to bear on the rules of the game rather than on the players, and finally in which there is an environmental type of intervention instead of the internal subjugation of individuals. I will try to develop some of all this next week.[64]

Unfortunately, Foucault did not keep his promise. He never cast his gaze into the future.[65] In the next sections, I will sketch out how the future looks like from today's perspective.

6 The Government of Modular Living Beings

Ivan Illich, in his last years, vehemently opposed the rise of a new artificial god, "a pseudo-god, and a negation of the God who took on flesh and who redeemed us."[1] As he pointed out, the proper name of this new pagan divinity is "life":

> Thinking about life makes us act as if there were life, although scientists never use the term, and no reasonable philosopher or ethicist would ever dare to introduce that term without much qualification into his argument. We are here before the emergence of some ultimate justification for letting ourselves be administered by a clergy, a managing clergy, a planning clergy, a dictatorial clergy, worse than anything we have ever thought about.[2]

With hindsight, particularly after the COVID-19 pandemic, Illich's words can be deemed prophetic. Whatever one's take on the sanitary crisis might be, it is a fact that a "managing clergy" has taken control of our lives all over the world.

In the following, I will focus again on the pseudo-god in the name of which this managing clergy administers the human population, and I will discuss one of its most curious epiphanies: Artificial Life. However counterintuitive this notion may seem, not only does it make perfect sense for information technology and computer science researchers, but it also sheds light on some crucial assumptions lying at the heart of both the neo-Darwinian view of Life and the behavioristic theory of the human mind. As will be shown, such assumptions lay the groundwork for a new type of biopolitical management of the human population that makes us shift from the government of the living to the government of modular living beings.[3]

6.1 Population of Entities

Although Illich had reason to denounce the growing prominence of what he called a new "managing clergy," he was wrong in saying that scientists never speak or think about Life as such. Life is, in fact, the theoretical construct, or the *abstraction*, in which modern evolutionary biology is grounded. What does it mean to be alive? For neo-Darwinian scientists, the answer is:

> We shall regard as *alive* any population of entities which has the properties of multiplication, heredity and variation. The justification for this definition is as follows: any population with these properties will evolve by natural selection.[4]

In this passage, the key concepts are "natural selection" and "population of entities," which both contribute to clarifying the meaning of Life from a neo-Darwinian point of view. The two concepts are closely connected. Indeed, as John Maynard Smith explains, only populations of entities can evolve by natural selection, thereby enhancing their survival capabilities down through the generations. Therefore, only populations of entities can be thought of as alive according to the modern evolutionary synthesis. Importantly, moreover, such populations of entities are not to be seen as sets of living beings, or organisms. Rather, they are to be seen as sets of adaptive traits that can be considered one by one and have distinct evolutionary stories. For neo-Darwinian biologists, in other words, it is not the individual organism that can be said to be alive in the first place, because it is on adaptive traits rather than individual organisms that natural selection acts. These traits are the entities of which populations are composed, and it is through them that Life flows, carrying evolution forward. As Stephen Jay Gould and Richard Lewontin recall in their manifesto against the neo-Darwinian "adaptationist programme":

> An adaptationist programme has dominated evolutionary thought in England and the United States during the past 40 years. It is based on faith in the power of natural selection as an optimizing agent. It proceeds by breaking an organism into unitary "traits" and proposing an adaptive story for each considered separately. Trade-offs among competing selective demands exert the only brake upon perfection; non-optimality is thereby rendered as a result of adaptation as well.[5]

I will return to natural selection as an optimizing agent later in this discussion. For the time being, it is worth emphasizing that, by considering adaptive traits separately, the neo-Darwinian adaptationist programme drops the assumption that individual organisms constitute "integrated wholes," as Gould and Lewontin remark.[6] The populations of entities brought to light by the neo-Darwinian theory of Life are populations of independent unitary traits rather than of integrated organisms. And it can already be noted that this idea is exactly the same one that permeates the new biopolitical technorationality of our time. In both cases, living beings are treated as modular assemblages of elements with no essential cohesion or intrinsic architecture.

6.2 The Axiomatic Procedure

At the very first stages of Artificial Life research, John von Neumann tackled the question of what makes us alive along similar lines. For him, to be alive means first and foremost to be able to reproduce and propagate, whatever the medium of Life may be. With regard to the organisms that can be found in nature, they "can be viewed as made up of parts which to a certain extent are independent, elementary units."[7] Therefore, when one studies natural systems such as living organisms, one can break the problem into smaller parts. One may "view as the first part of the problem the structure and functioning of such elementary units individually. The second part of the problem consists of understanding how these elements are organized into a whole, and how the functioning of the whole is expressed in terms of these elements."[8] In short, von Neumann advocates a bottom-up approach to the study of living beings: first, we identify the elementary units of which living beings are made, and we describe their functioning; then, we investigate how these elements are combined into a whole. To achieve the first task, he suggests adopting an "axiomatic procedure." For him, it is not necessary to analyze in depth the elementary units of which organisms are composed, exploring in detail their organic and physical chemistry. We can limit ourselves to extrapolating their basic characteristics from their specific behaviors:

> *The Axiomatic Procedure.* Axiomatizing the behavior of the elements means this: We assume that the elements have certain well-defined, outside, functional characteristics; that is, they are to be treated as "black boxes." They are viewed as automatisms, the inner structure of which need not be disclosed, but which are assumed to react to certain unambiguously defined stimuli, by certain unambiguously defined responses.
>
> This being understood, we may then investigate the larger organisms that can be built up from these elements, their structure, their functioning, the connections between the elements, and the general theoretical regularities that may be detectable in the complex syntheses of the organisms in question.[9]

In the essay from which this quotation is taken, von Neumann declares that he is aware that this method is far from satisfactory from an epistemological point of view, but he nonetheless deems that the axiomatic procedure allows researchers to cut to the chase and focus directly on the most relevant "automatism" that characterizes biological processes in general: self-reproduction. Once this automatism has been analyzed and understood in logical and mathematical terms, it becomes possible to conceive of artificial automata that somehow imitate natural organisms and can help us to improve our interpretation of biological living systems. Unsurprisingly, given the method's limitations, the final result—a theory of self-reproducing automata—will be nothing more than an approximation, as von Neumann emphasizes. This means that, according to him, Artificial Life is an *abstraction*, a conceptual model of the living (or, more precisely, of the gene) that is not perfectly identical, but only "similar to reality."[10]

However, if we now return to the neo-Darwinian research programme, we can notice that von Neumann's abstraction is not so different from what modern evolutionary theorists consider to be the *reality* of Life. Indeed, living beings as viewed from a neo-Darwinian perspective are assemblages of elementary units, adaptive traits. The only difference between neo-Darwinian biologists and von Neumann is that the former do not think of these elementary units as the building blocks of organisms. Rather, they see them as the elements of a higher-order whole, the population, which really is alive, reproduces, and evolves by means of natural selection. That said, the basic assumption remains the same: living beings are aggregates of elementary units.

Moreover, for both von Neumann and neo-Darwinian biologists, such elementary units are identified by behaviors. Following von Neumann's axiomatic procedure, the elementary units of the living are identified by certain unambiguously defined responses to certain unambiguously defined stimuli. When it comes to the neo-Darwinian research programme, adaptive traits are identified by the behaviors that they enable living beings to adopt—for instance, claws are to be considered adaptive traits insofar as they enable living beings to *do something* (to dig, to climb, to grasp prey, and so forth). Thus, whether we speak the language of Artificial Life or the language of neo-Darwinian biology, living beings turn out to be aggregates of behaviors.

6.3 Radical Behaviorism

The critical role that the category of behavior plays in the theory of natural selection was clearly acknowledged by one of the most important exponents of behaviorism, B. F. Skinner, the father of radical behaviorism, who gave the following definition of the "behavior of organisms":

> Behavior is what an organism is *doing*. . . . Behavior is that part of the functioning of an organism which is engaged in acting upon or having commerce with the outside world. . . .
>
> By behavior, then, I mean simply the movement of an organism or of its parts in a frame of reference provided by the organism itself or by various external objects or fields of force. It is convenient to speak of this as the action of the organism upon the outside world, and it is often desirable to deal with an effect rather than with the movement itself, as in the case of the production of sounds.[11]

According to Skinner, this definition is key to the study of both human and nonhuman living beings. There is nothing special about human behavior compared to nonhuman behavior; all organisms move and interact with the outside world, and therefore, all organisms behave. On this point, there seems to be no significant difference between Skinner and von Neumann: both think that Life finds expression in behavior.

The similarities between the two approaches do not stop here, for both von Neumann and Skinner are "black box" thinkers. The axiomatic procedure of von Neumann generates "black boxes," as he says, in that it deliberately leaves out of sight what hides behind the living's behaviors (namely, the "inner structure" of the living's elementary units or their "organic and physical chemistry"). For von Neumann, we need not investigate what is concealed within these "black boxes" to comprehend and compute the automatism that characterizes the Life of living beings—self-reproduction. Hence the famous definition attributed to him, but for which no source can be found: "Life is a process which can be abstracted away from any particular medium."[12] Whether or not von Neumann ever uttered these words, they do sum up his thoughts on the matter.

Skinner is even more radical. In his psychological investigations, he reaches the conclusion that the mind is a black box, but for him, this amounts to saying that the box itself is a mirage: it has no internal unity or permanent identity. In reality, the mind is nothing but an aggregate of discrete behavioral patterns that can be observed and studied in broad daylight. Thus, to understand what we are talking about when we talk about the mind, all we need to do is break the human mind into separate behavioral units, just as von Neumann breaks the organism into distinct elementary units and neo-Darwinian biologists break the organism into unitary adaptive traits. In this manner, a process of desubstantialization—of the mind in one case, of living beings in the other—is started. And Skinner's radical behaviorism helps us to grasp some radical, yet often-unnoticed implications of the process of desubstantialization carried out by the neo-Darwinian research programme.

Every time that natural selection is taken into consideration, the question arises as to what the object of selection is. On what kind of *substance* does natural selection act? Ever since Darwin introduced the theory of evolution by means of natural selection, this question has been at the center of heated debates, and today, there is still no agreement among neo-Darwinian biologists on the topic. Some say that the "units" of selection are the genes, others reply that the "target" of selection is the phenotype, and yet others contend that natural selection can also operate at the level of groups. Even though the gene-centered view has been prevalent among evolutionary biologists over the past century, it seems that neither of the aforementioned possibilities can be a priori excluded.[13]

Skinner touched on this topic, cursorily, during a conversation with Edward O. Wilson at Harvard in November 1987. For Wilson, the father of sociobiology and a strong proponent of the neo-Darwinian approach to evolution, natural selection develops on two levels: genes and cultural groups. On both levels, according to Wilson, population of entities go through a process of selection—whence his late theory of "culturgens" and "gene-culture coevolution."[14] It might be objected that this theory, coauthored with Charles Lumsden, continues to rely heavily on the notion of genetic transmission and somehow restates a gene-centered view of evolution,[15] but Wilson nonetheless considered cultural variation a crucial ingredient in the evolution of humans. There remained the problem of the fundamental unit or units of selection. And this is where Skinner enters the scene: "I think that is something very few people properly understand; that is, the difference between the causality we can observe in physics and the selection action on the behavior of living things."[16]

During his conversation with Wilson, Skinner stresses that the human mind cannot be analyzed starting from the genes, or the nervous system, or whatever material substrate. The mind is a black box, he maintains, in that there is nothing in it other than behaviors, which are the basic units of the mind and cannot be reduced to anything else. To explain behaviors, one has to start from other behaviors, which constitute the nonsubstantial matter, as it were, studied by psychologists. And various kinds of selection, from natural to operant, can produce changes in the living's behaviors, as Skinner points out:

> I'm a "black box" man myself. I think I was the first behaviorist to say I don't care about what's going on in the mind and I don't care what's going on in the nervous system. . . .

> You must have reinforcing consequences at every stage. You just can't simply say, oh, well, there's some pattern of genes and they just simply occurred. . . .
>
> It is not merely cause and effect of the push-pull, stimulus-response sort, because selection comes in at all three levels.[17]

Here, the three levels of selection to which Skinner alludes are presumably natural selection, individual operant selection, and group operant selection, as the editor of the dialogue suggests. I will go over the difference between operant and natural selection shortly. Before that, though, we can already read between the lines the idea on which Skinner had been working for years. If selection in general (now natural, now operant) comes in at several levels, this is because selection, whether natural or operant, is not like a physical cause that acts on a certain substratum or medium through "push and pull." Rather, selection acts on "the behaviors of living things," which can be analyzed without taking into account the black box of their material substrata. Clearly, this leaves open the question of how selected behaviors are transmitted from one generation to the next. To answer this question, one needs to examine the particular nature of the populations that undergo selection. But however debatable Skinner's thesis may seem for this and other reasons, one cannot deny that he found a way around the problem of deciding what the ultimate level (object, unit, or target) of selective processes is—or, to put it in slightly different terms, what the true medium of Life is.

With regard to natural selection, if we assume that it acts on behaviors rather than anything else, it follows that it can occur at the level of both genes and cultural groups, as Wilson believed, because on both levels, there will be behaviors from which to select: everything that is alive behaves in one way or another. A gene expresses a behavior, a group expresses a behavior. And the same holds true for a phenotypic trait. Moreover, as already explained, scientists can actually recognize a gene, an adaptive trait, or a cultural group only by the behaviors that each of these entities manifests. Thus, each of these entities becomes relevant to scientists only so long as each of them expresses a certain behavior that will eventually evolve into another, entailing the activation of a particular type of inheritance mechanism. For Skinner, the conclusion to be drawn from all this is that the process of selection can be abstracted from all these entities, for selection ultimately concerns the evolution of behaviors themselves, regardless of what the medium or the carrier of behavior is. In the end, evolution by means of selection is a behavioral process. Therein lies the secret of Darwin's revolution: Life is a process that can be abstracted away from any particular medium.

6.4 Behavioral Selectionism

Again, one is struck by the similarities between Skinner's view and von Neumann's approach to the question of Life. And it is worth recalling at this point that Skinner made explicit reference to computer science research during his conversation with Wilson:

> Al Newell, who was here to give the William James lectures, said he and Herb Simon have computer systems now worked out that do learn operant behavior. That is, the consequences changed them so that they behave in a given way. This is perfectly feasible. After all, here we do this. I don't see why a machine can't be made to do it.[18]

This remark shows that Skinner took the idea of strong Artificial Life very seriously. According to many researchers, if machines can synthesize, rather than merely simulate, the behavior of intelligent beings that are capable of learning, then there is no reason to exclude that machines can be made to synthesize, rather than merely simulate, the behavior of living beings that are capable of self-reproduction. In the final analysis, if it all comes down to behavior when we analyze the living's interactions with the outside world, it seems more than plausible that machines too can play the game of Life, provided that they learn to behave as living beings. This is what Christopher Langton, a prominent figure in Artificial Life research, had claimed just two months before Skinner and Wilson's conversation, at the first workshop on Artificial Life held in Los Alamos in September 1987:

> Artificial Life is the study of man-made systems that exhibit behaviors characteristic of natural living systems. It complements the traditional biological sciences concerned with the *analysis* of living organisms by attempting to *synthesize* life-like behaviors within computers and other artificial media. . . .
>
> The "key" concept in AL is *emergent behavior*. Natural life emerges from out of the organized interactions of a great number of non-living molecules, with no global controller responsible for the behavior of every part. Rather, every part is a behavior itself, and life is the behavior that emerges from out of all of the local interactions among individual behaviors.[19]

As this much-quoted passage attests, strong Artificial Life shares some basic assumptions with radical behaviorism, in particular the idea that behavior is the stuff of which Life is made. But what about selection?

For Skinner, there is a close connection between the two concepts, in that the "emergent behavior" of living beings—to put it in Langton's words—is constantly subject to selection. In a nutshell, Skinner's selection can be described as the impersonal action of Mother Nature that chooses between patterns of behavior that first appear and then become consolidated or disappear because of the consequences, either beneficial or detrimental, that follow. Skinner calls this process "selection by consequences" and distinguishes between two types of selection. One is "natural selection," which thus turns into a particular instance of a more general class of natural phenomena; the other is "operant conditioning," a concept that probably represents Skinner's major contribution to psychology. Natural selection of behavior is pushed forward by favorable consequences that are termed "contingencies of survival." As for operant conditioning, it hinges on "contingencies of reinforcement." This is how Skinner himself presented the idea of selection by consequences in 1981:

> Human behavior is the joint product of (i) contingencies of survival responsible for natural selection and (ii) contingencies of reinforcement responsible for the repertoires of individuals, including (iii) the special contingencies maintained by an evolved social environment. Selection by consequences is a causal mode found only in living things, or in machines made by living things. It was first recognized in natural selection. Reproduction, a first consequence, led to the evolution of cells, organs, and organisms reproducing themselves under increasingly diverse conditions. The behavior functioned well, however, only under conditions similar to those under which it was selected.
>
> Reproduction under a wider range of consequences became possible with the evolution of processes through which organisms acquired behavior appropriate to novel environments. One of these, operant conditioning, is a second kind of selection by consequences. New responses could be strengthened by events which followed them. When the selecting consequences are the same, operant conditioning and natural selection work together redundantly. But because a species which quickly acquires behavior appropriate to an environment has less need for an innate repertoire, operant conditioning could replace as well as supplement the natural selection of behavior.
>
> Social behavior is within easy range of natural selection, because other members are one of the most stable features of the environment of a species. The human species presumably became more social when its vocal musculature came under operant control. Verbal behavior greatly increased the importance of a third kind of selection by consequences, the evolution of social environments or cultures. The effect on the group, and not the reinforcing consequences for individual members, is responsible for the evolution of culture.[20]

Operant conditioning—unlike respondent conditioning (also known as "classical conditioning" or "Pavlovian conditioning"), by means of which living beings learn to pair a biological stimulus (e.g., food) with a previously neutral stimulus (e.g., a bell)—is the way that living beings selects new behaviors. For this to be possible, new behaviors must be "emitted," Skinner avers.[21] This emission of behaviors plays more or less the same role that variation plays in natural selection: it generates a variety of behaviors from which it is possible to select one. According to Skinner, the capacity to emit behaviors is in itself a behavior produced by natural selection, a kind of adaptive metabehavior (my definition) that allows living beings to invent new patterns of behavior, which then can be subject to operant conditioning. After new behaviors have been emitted, a selection is made on the grounds of the consequences that such behaviors have. If the consequences are good, behaviors will be reinforced automatically; otherwise, they will be expunged from the behavioral repertoire of the organism. Living beings take no initiative in this process; they are not the agents of selection. Quite the contrary, in fact—Skinner claims that they are the product of this process: an assemblage of behaviors. In other words, there is no mind or internal agency that decides which behaviors are beneficial and which not, because the mind itself is just an aggregate of behaviors that arises from, and is constantly modified by, the process of selection by consequences. For Skinner, the whole (the mind, the living) is *not* greater than the sum of the parts (behaviors).

Many aspects of this highly speculative theory deserve close attention. The first thing to note is that selection by consequences is a "causal mode" that can be found only in living beings and, Skinner adds, "machines made by living things." Thus, the hypothesis of strong Artificial Life gains impetus: since selection by consequences is the "causal mode" that explains both the evolution of living beings by means of natural selection (one type of selection by consequences) and the operant behavior of living beings and learning machines made by living beings (the other type of selection by consequences), the boundaries between natural and artificial become blurred. In the end, if Life can be abstracted away from any particular medium, then any medium can be transformed into something alive by those who know the ultimate secret of Life, selection by consequences. And those who know this secret will also be able to breathe new Life into living things if the need arises. The former is the task of the Artificial Life engineer; the latter of the behavioral therapist.

Indeed, the mission of the behaviorial psychologist is precisely to infuse Life into the living and the whole of society, according to Skinner. And his theory has momentous practical consequences, as will be shown shortly. Yet, before expanding on this aspect of radical behaviorism that gives us a hint about how the biopolitical management of behavior is understood and practiced today, it is worth highlighting that the notion of selection by consequences also allows Skinner to tackle a notorious problem in evolutionary biology: teleology. Skinner is persuaded that he can solve this problem once and for all.

The alleged solution is provided by the concept of consequence, which is meant to replace the concept of purpose, thereby clearing the field of all teleological readings of selective processes. Take, for example, the mechanism of operant conditioning. In this case, behaviors are selected on the basis of their effects within the environments in which they occur. If the effects are beneficial, behaviors will be reinforced and reiterated; otherwise, they will disappear. Since the effects of behaviors come after the behaviors, it may seem that some sort of teleological reasoning vitiates this argument: the effect of a certain behavior (observed at time = 1) should *explain* something prior to it (observed at time = 0).[22] But since Skinner tests this conjecture on animals (pigeons and rats) lacking any "free will" and succeeds in modifying their behavior by means of operant conditioning, he becomes convinced that selection by consequences is a causal mode that does not involve any kind of purpose, purposiveness, or goal-directed automatism, and this type of causation is purely deterministic. Given this premise, he then generalizes the notion of selection by consequences and describes the emergence of evolutionary traits along the same lines. Adaptations—or more accurately, adaptive behaviors—are selected because of the beneficial consequences they have, not because they serve any purpose. Natural selection is just a particular instance of selection by consequences.

Things are not that simple, however. With regard to operant conditioning, it might be objected that Skinner's explanation, albeit intuitively powerful, is by no means an explanation. In fact, for a selection to be possible, there has to be a preliminary comparison between two or more sequences of events, as well as an evaluation of the best outcome, no matter how primitive this operation of a cognitive nature might be in animals such as pigeons or rats. In other words, even though pigeons and rats cannot abstract the notion of purpose from their behavior, even though they cannot reflect on what a purpose is in and of itself, it seems mandatory to posit that they are at least capable of pondering over alternative sequences of events and perceiving which result is the most convenient for them to achieve when their behavior is selected on the basis of its effects or consequences. Otherwise, it would be altogether inappropriate to even speak of a selection by *consequences*. Who sees that B is a *consequence* of A? If it is said that the psychologist and nobody else sees this, then the word "consequence" becomes merely descriptive. If we hold, on the contrary, that the animal sees this, then the word is explanatory, but the explanation in turn is mentalistic.[23]

This objection may seem powerful, and yet it is not decisive because the problem looks different to a radical behaviorist like Skinner. It is not the living being that chooses the most beneficial options by comparing the consequences of various behaviors, since there is no mind that conceives or makes a mental representation of these behaviors. Here, behavior is all that exists. But then, how are behaviors selected? How are the effects of behaviors assessed? The answer is twofold. On the one hand, selection by consequences is achieved through natural selection; in this case, selected behaviors are those adaptive behaviors that ensure survival, as the neo-Darwinian theory of evolution says. On the other, given that Skinner does not endorse an ultrareductionist view of the living, according to which everything is predetermined by evolution, he also claims that not all behaviors are innate and can be traced to the action of natural selection. This is where the science of psychology comes to the fore. If all behaviors were predetermined by natural selection, there would be no room for a psychological interpretation of some of them: everything could be explained in biological or sociobiological terms. But this is not the case, as Skinner objects to Wilson.[24]

Even when the living's behaviors are not determined by natural selection, however, Skinner thinks that they are guided by the same general law, selection by consequences, of which natural selection is a special instantiation, and operant conditioning another. According to him, therefore, it can be assumed that some behaviors have arisen from natural selection and passed the test of the "contingencies of survival," whereas others have arisen from operant conditioning and passed the test of the "contingencies of reinforcement." In both cases, the selected behaviors are necessarily the most beneficial to the living. And this means that they always lead to the same consequence: survival.

Thus, teleology reappears within Skinner's theoretical framework, despite all efforts to eradicate it from scientific reasoning, because survival turns out to be the omnipresent telos—renamed "consequence"—that steers both types of selection by consequences.[25] This tacit assumption, problematic as it may be, is nonetheless necessary in order to deny that the mind has an independent existence and a controlling function. In Skinner's view, "the mind is behavior" entails that behaviors emerge from a process, selection by consequences, of which living beings are the combinatorial product rather than the cause. Here, living beings are the results, not the sources, of their own behaviors. Clearly, the living's behaviors cannot be thought of as purely mechanical movements because living things are different from nonliving things and lifeless matter. But if behaviors lend themselves to psychological interpretation and have a more-than-mechanistic meaning, this is not because they are guided by an invisible mind, but rather because all of them arise from a process that always points in the same direction and leads to survival. In this way, behaviors—unlike the movements of the planets around the sun—reveal themselves to be the expression of a *will to survive* that defines Life as such, or Life abstracted away from any particular living system. From Skinner's perspective, such a goal-oriented, yet unintentional will is the distinctive trait of all living things, from nonhumans to humans, including living machines built by humans:

> All human behavior, including the behavior of the machines which man builds to behave in his place, is ultimately to be accounted for in terms of the phylogenic contingencies of survival which have produced man as a species and the ontogenic contingencies of reinforcement which have produced him as an individual.[26]

6.5 A Technology of Behavior

Skinner's reference to machines is not fortuitous because "man *is* a machine" rather than an intentional agent.[27] More generally, all living beings can be thought of as living machines, insofar as their behavior undergoes a process of selection by consequences. And the same holds for cultures because cultural behavior too—like the behavior of organism and the behavior of species—goes through a process of selection that constantly increases its survival value. Accordingly, selection occurs at three levels from Skinner's point of view. At the level of species, an increase in survival value is achieved through natural selection. At the level of organisms, a similar increase in value is achieved through operant conditioning, which improves the living's "susceptibility" and capacity to adjust to rapidly changing environments—a capacity, highly developed in humans, that is in turn a product of natural selection. At the level of cultures, finally, an increase in value is achieved every time cultural patterns strengthen social behaviors that favor survival. For this reason, Skinner proclaims that the science of behavior, a branch of the life sciences, is nothing less than "a science of values."[28]

The axiological characterization of the science of behavior is a direct consequence of its teleological assumptions. If behaviors naturally aim for survival, then survival is that which ultimately gives value to all behaviors and, by the same token, to all values expressed by human behaviors over the centuries. Skinner thus expands Darwin's interpretation of survival in terms of a nomological/axiological principle.[29] In doing so, he adopts the same perspective as his highly esteemed predecessor: the axiological assessment of behaviors made possible by the science of behavior does not entail that living beings are free to choose between good and bad behaviors because bad behaviors go automatically extinct, whereas good ones necessarily persist and survive, regardless of the living's intentions. From this view, the history of human behaviors, just like the evolution of Life in general, is governed by strictly deterministic laws that dictate the course of events and lead to a better future, no matter how humans see things from within the limited perspective of their own cultures and societies. As a result, behavioral scientists like Skinner do not confine themselves to arguing that this or that value is morally, culturally, or socially preferable, as ethicists could possibly do. Rather, Skinner measures the value of *all* cultural values against the survival value that *each* of them has. Thus, he finds himself in a position to predict the direction that human history will inevitably follow. The science of behavior understood as "a science of values" is also a philosophy of history:

A culture, like a species, is selected by its adaptation to an environment: to the extent that it helps its members to get what they need and avoid what is dangerous, it helps them to survive. . . .

New practices correspond to genetic mutations. A new practice may weaken a culture—for example, by leading to an unnecessary consumption of resources or by impairing the health of its members—or strengthen it—for instance, by helping members to make a more effective use of resources or improve their health. . . .

Just as we do not need to explain the origin of a genetic mutation in order to account for its effect in natural selection, so we do not need to explain the origin of a cultural practice in order to account for its contribution to the survival of a culture. The simple fact is that a culture which *for any reason* induces its members to work for its survival, or for the survival of some of its practices, is more likely to survive. Survival is the only value according to which a culture is eventually to be judged, and any practice that furthers survival has survival value by definition.[30]

For these reasons, Skinners argues, the history of human cultures has a certain direction, which does not depend on what human beings think but only on the automatism of selection by consequences, which tends to maximize the survival value of everything that is alive.[31] The history of culture follows the same laws that govern the history of nature. The living aim for survival: this is the Darwinian axiom on which Skinner grounds his behavioristic philosophy of history. In light of all this, we can start to understand, on the one hand, what future awaits us in Skinner's view, and on the other, how close to present-day reality his prediction is. The closeness is such that one might even wonder whether Skinner's behavioral science has directly influenced the course of history to a much greater extent than is generally acknowledged. Here, however, I will not pay close attention to this issue, nor will I examine in detail the many interpretations of Skinner's theory and various elaborations of its concepts that have exerted a significant impact on the development of the social sciences over the last decades. I will just take Skinner as a visionary, as he liked to portray himself: the prophet of a brave new world, the designer of a utopian society, the evangelist of a new biopolitical regime.[32]

In brief, the future that Skinner sees coming is distinguished by five characteristics: control, optimization, modularity, depersonalization, and disembodiment. These characteristics, which I will discuss one by one next, are key to understanding both the way that Skinner conceives of human society and the way that the human population is governed nowadays.

Control is the concept around which Skinner's theory centers. The idea is that one's behavior can be commanded and modified by others, and this operation can be achieved through control of the environment in which humans live and through manipulation of the contingencies of reinforcement that the environment provides. "Behavior can be changed by changing the conditions of which it is a function."[33] A "technology of behavior," Skinner's definition of the technology of power that the science of behavior brings forth, consists in changing people's behaviors by changing the environmental responses to them.

As mental health professionals know, this is what is ordinarily done by behavioral psychotherapists, who use the technique of operant conditioning to treat their patients. Yet Skinner's technology of behavior control does not apply solely to the field of psychotherapy. Indeed, his thesis is that to treat mental disorders and to govern human beings basically mean the same thing—namely, to modify people's behavior without asking them what they consider to be good, or at least better, for themselves and others. The therapist knows what is truly good, and so should the politician: good is anything that furthers survival. The phrase "technology of behavior" refers to this particular type of scientific knowledge and biopolitical technorationality, which, according to Skinner, should become hegemonic everywhere in the world to address the life-threatening problems with which contemporary societies are confronted.

There are a few things to note about the new technology of behavior control that Skinner regards as essential for the government of present-day and future societies. First, this technology is blatantly antidemocratic. Behavior control paves the way to heterodirect behavior, which is the opposite of self-determination. Furthermore, Skinner's technology does not even target people's behavior straightforwardly; rather, it manipulates the environments in which behaviors occur so as to cause new behaviors to appear and replace, almost automatically, the old ones by virtue of more beneficial consequences. Most of the time, this automatism does not require people to be aware of anything. In reality, Skinner avers, people never change behavior of their own free will, after reasoning on their own utility, feelings, or moral duty. This explains why he attacks the notion of "the autonomous man" and ridicules "the literatures of freedom and dignity":

> Life, liberty, and the pursuit of happiness are basic rights. But they are the rights of the individual and were listed as such at a time when the literatures of freedom and dignity were concerned with the aggrandizement of the individual. They have only a minor bearing on the survival of a culture. . . .
>
> Those who have been induced by their culture to act to further its survival through design must accept the fact that they are altering the conditions under which men live and, hence, engaging in the control of human behavior. Good government is as much a matter of the control of human behavior as bad, good incentive conditions as much as exploitation, good teaching as much as punitive drill. Nothing is to be gained by using a softer word.[34]

The second thing to note about Skinner's technology of power is the dramatization of the political agenda. Since the technology of behavior control operates by means of selection by consequences and is geared toward increasing the survival value of behavior, this technology can only discriminate between life-threatening behaviors, which may lead to extinction, and life-preserving behaviors, which will ensure survival or at least increase survival chances. Here too, Skinner finds himself in total agreement with Darwin, who already saw the variations upon which natural selection acts as being either deadly or salvific for the living.[35] In both cases, the selective process is thought of as a natural tragedy where life-threatening and life-preserving options confront each other on a constant basis. Thus, Skinner's technology of behavior control hinges on the assumption that society is constantly facing emergency situations that expose it to the risk of extinction:

> In trying to solve the terrifying problems that face us in the world today, we naturally turn to the things we do best. We play from strength, and our strength is science and technology. To contain a population explosion, we look for better methods of birth control. Threatened by a nuclear holocaust, we build bigger deterrent forces and anti-ballistic-missile systems. We try to stave off world famine with new foods and better ways of growing them. Improved sanitation and medicine will, we hope, control disease, better housing and transportation will solve the problems of the ghettos, and new ways of reducing or disposing of waste will stop the pollution of the environment. We can point to remarkable achievements in all these fields, and it is not surprising that we should try to extend them. But things grow steadily worse. . . .
>
> What we need is a technology of behaviour.[36]

The third thing to note about Skinner's theory and technology of power is the prominent role in the government of society that it assigns to scientists. For instance, we may wonder: Who is entitled to identify the "terrifying problems" that humanity is facing? For Skinner, the answer goes without saying: only scientists can be given the task of making such a diagnosis because only scientists know how to evaluate the nature and gravity of problems, and how to tackle them appropriately. Again, Skinner's antidemocratic inclination becomes apparent here, together with the vexing similarities between the technocratic elites of which he dreams in his writings and those that have taken control of Western societies over the last decades.[37] As a matter of fact, not only do the latter ground their decisions on the same emergency logic most of the time, but they are also becoming accustomed to implement such decisions through the same technology of power, as we will see shortly.

And yet, with few exceptions, critical thinkers have generally overlooked the impact of Skinner's radical behaviorism on the evolution of biopolitical power. Perhaps the reason for this is that Skinner's arguments about the necessity of applying his technology of behavior on a global scale to avoid the extinction of the human race are so mind-blowing and dystopian that they seemed totally unrealistic to most people when they were made public for the first time.[38] However, Skinner's notion of reinforcement in operant conditioning, just to give an example, continues to be highly influential these days, integral as it is to both the cognitive-behavioral treatment of mental disorders and the managerial-behavioral approach to the government of the human population.

According to Skinner, reinforcement can be negative or positive: negative reinforcement consists in promoting a certain behavior by removing an aversive stimulus, whereas positive reinforcement consists in promoting a certain behavior by modifying the environment in such a manner that a beneficial consequence may follow from that behavior. Today, both techniques find wide application in several domains of economic, social, and political management, where they sometimes reappear under a different name, such as "nudging."[39] Although Richard Thaler and Cass Sunstein do not speak of "reinforcement" in their seminal work on the topic, positive reinforcement in Skinner's sense of the concept is essential to their theory of behavioral nudges, which has become a crucial ingredient of the science of social engineering over the last dozen years.[40] Likewise, positive reinforcement lies at the heart of the new technology of power commonly known as the "social credit system," first introduced in China and then exported to Western countries.[41] To cut a long story short, positive reinforcement is among the most used and important instruments of biopolitical power nowadays. It is to be hoped that this will become a topic of interest to students of biopolitics in the years to come.

As Noam Chomsky has pointed out recently, despite the fact that the notion of behavioral reinforcement makes little sense from the point of view of rigorous science, and despite the fact that it has been refuted countless times in the past, this notion and the theory behind it are "right back the next day."[42] In all likelihood, the main reason for this is that the concept of behavioral reinforcement represents one of the pillars of today's biopolitical technorationality. Drawing on this and other concepts originally coined by Skinner, applied behavioral research has developed hugely since the 1970s and 1980s and has heavily influenced the way that biopolitical control is conceived and practiced these days—so much so that partnerships between Behavioural Insights Units and national governments are now multiplying all over the world (e.g., in the UK, the US, Australia, New Zealand, France, and many other countries). These partnerships show, among other things, that the biopolitical reinforcement of behaviors and the sovereign enforcement of laws are likely to become almost indistinguishable as time goes by. As Shoshana Zuboff, a former student of Skinner, has underscored:

> "Conditioning" is a well-known approach to inducing behavior change, primarily associated with the famous Harvard behaviorist B. F. Skinner. He argued that behavior modification should mimic the evolutionary process, in which naturally occurring behaviors are "selected" for success by environmental conditions. Instead of the earlier, more simplistic model of stimulus/response, associated with behaviorists such as Watson and Pavlov, Skinner interpolated a third variable: "reinforcement." . . .
>
> Skinner called the application of reinforcements to shape specific behaviors "operant conditioning." His larger project was known as "behavior modification" or "behavioral engineering," in which behavior is continuously shaped to amplify some actions at the expense of others. . . .

As the chief data scientist for a much-admired Silicon Valley education company told me, "Conditioning at scale is essential to the new science of massively engineered human behavior."[43]

The influence of Skinner's thought in the field of education sciences has attracted some attention lately,[44] but much remains to be done to fully appreciate Skinner's contribution to the development of "the new science of massively engineered human behavior." Having said that, it must also be emphasized that the true importance of Skinner lies not in this or that concept but rather in his overall view of human beings and society, a view from which the new biopolitical regime of our time seems to draw inspiration in its daily exercise of power and control over the population. The necessity and modalities of such an invasive control were clearly explained for the first time in history by Skinner, the visionary who perceived well ahead of his time the future that awaited us. In some respects, one might even say that today, Skinner can be attributed more or less the same role that Michel Foucault assigned to Jeremy Bentham in his pioneering works on disciplinary societies.[45] Both Bentham and Skinner provide us with a simplified model of how biopower functions and shapes human beings at a particular moment in history: in contemporary societies, the disciplinary protocols and apparatuses of normalization are giving way to the new ethopolitical protocols and apparatuses of control; thus, Bentham's panopticon is now being superseded by the Skinner box, the miniature of the world to come.

6.6 Optimization and Modularity

The Skinner box is a fascinating device. It is a cage where researchers can observe and study an animal's response to classical and operant conditioning, with a view to understanding how to manipulate its behavior. In many regards, the Skinner box resembles Bentham's panopticon, for those who are observed cannot see those who observe them. There is, however, a significant difference between the two devices: Bentham's panopticon served to monitor individuals, whereas the Skinner box serves to monitor and manipulate behavior—which is why a direct equation between human beings and animals can be made in this case: the laws of behavior are exactly the same in either case.

As already said, this equation between animal and human behavior, as well as the theory that justifies the equation, have been called into doubt several times and do not appear credible.[46] Even though Skinner's technology of power seemingly succeeds in conditioning human behavior every now and then, it does not follow that the theory is correct, not least because Skinner's account of human behavior is oversimplistic and leaves too many facets of human existence unexplained. In the distant future, therefore, it is highly likely that radical behaviorism will be consigned to oblivion, suffering the same fate as the theories of normal and deviant individuals that were in vogue in the late nineteenth and early twentieth centuries. That being said, just as those theories reflected and partly determined how people saw themselves back in those days, so Skinner's theory tells us how people are learning to see themselves nowadays under pressure from the new biopolitical regime that Skinner himself has partly contributed to bringing into being.

Two implications of Skinner's theory are particularly relevant and illuminating: the amnesia that strikes all living beings, and the parcelization of individuals (which is also key to the neo-Darwinian research programme). Such implications are the theoretical counterpart to two basic precepts of the new biopolitical regime: optimization and modularity.

For a radical behaviorist, operant conditioning involves amnesia because no amount of information is stored inside the mind during the process of conditioning.[47] As a result of reinforcement, one behavior is replaced with another, without leaving any detectable trace of this process inside the black box. The present cancels the past, and that is all. In Skinner's opinion, there is no reason to think otherwise. The mind, viewed as a "behavioral repertoire," is reshaped through operant conditioning, but no memory of past behaviors remains unless conditioning has failed. Hence, Skinner argues, psychologists and cognitive scientists should stop talking about mental representations or exchanges of information between the living and the environment. They had better concede instead that a change in behavioral patterns is directly, almost mechanically, induced by a change in the environmental conditions.[48]

For operant conditioning to be possible, however, some sort of assessment of what is good, or better, must be presupposed—otherwise, selection by consequences would turn out to be a neutral process that leads nowhere. In the final analysis, it is precisely this tacit assessment that differentiates living beings from lifeless machines: good is anything that furthers survival, and living beings are machines that are characterized by an unconscious awareness of what is "good" and what is "bad." But how should we think of any awareness here? As a matter of fact, if living beings were aware of anything, if they were able to discern what is better for them from a first-person perspective, it would be mandatory to conclude that they control their own behavior, that they are inhabited by a hidden homunculus, the "mind," that can evaluate the pros and cons of every behavior.

That idea is anathema to Skinner. As he writes, "science does not dehumanize man, it de-homunculizes him."[49] For the behavioral scientist, the self is not a supervising agent with an internal, substantial unity, but only "a repertoire of behaviour appropriate to a given set of contingencies."[50] Therefore, it is not the living being that can take the initiative and decide what is to be done on the basis of some acquired knowledge or previously stored information. Skinner's selection by consequences, like Darwin's natural selection, abides by an altogether different principle: it is Life as such, or Life in the third person, that pushes living beings to maximize the survival value of behaviors. Behaviors constantly increase their survival value not because they are controlled and selected by the living, but because it is Life itself that selects them through an autotelic process. This unspoken assumption lies behind Skinner's theory of human behavior, as well as Darwin's view of Life. And it is not the only important point of agreement between them.

Since the best possible behaviors are those that permit living beings to face the greatest variety of environmental conditions, optimal behaviors are those that express the greatest adaptability and capacity to adjust to new environments. Yet, as Darwin had already realized, adaptability and continuous adjustment do not imply that living beings can attain any state of perfect or normal equilibrium. According to Skinner, neither perfection nor normality belongs in this world because the evolution of life forms by means of natural selection and the evolution of behaviors through operant conditioning will go on forever and ever. Natural selection and operant conditioning, the two types of selection by consequences, are eternal laws of nature that will always govern the universe of living beings and never allow them to find rest. It follows that, first, *optimal* behaviors are not to be understood as *normal* behaviors because the natural world is not built on norms that guide its evolution and ensure its stability; and second, living beings do not possess any intrinsic unity but rather are to be seen as *modular machines* because behaviors change one by one and reshape living beings step by step, without pause. Thus, just as the organism represents an assemblage of adaptive traits from the point of view of the neo-Darwinian biologist, so the mind represents an assemblage of behaviors, or a "behavioral repertoire," from the point of view of the behavioral psychologist.

The concept of modularity turns out to be crucial in both cases. And modularity, in turn, entails that Life is a process that can be abstracted from any particular medium because Life is a process of optimization of living machines that does not coincide with any of them but rather forces all of them to change over time.

The difference between normalization and optimization reappears at this point: to normalize means to regiment the life and conduct of living beings according to some preestablished norms that are taken—or better, mistaken—for natural standards or patterns. On the other hand, to optimize means to maximize the survival value of life forms and behaviors without following any predetermined direction. In this case, no choice can be a priori excluded, provided that it maximizes survival value.

Far from being confined to the domain of biological and psychological research, the notion of optimization qua survival value maximization plays a fundamental role in many other fields of social science and has become the conceptual cornerstone of the new biopolitical-ethopolitical regime.[51] For example, it lies at the heart of the economic approach to human behavior propounded by Gary Becker, who may be considered one of the most important figures in twentieth-century social research.[52]

Although Becker made it clear that he did not want to associate with Skinner,[53] he agreed with the idea that all human behavior has a biological meaning, and he reached the conclusion that the study of economic behavior is connected to the study of behavior from a biological standpoint. But Becker went even further. In his view, economic theory can contribute to a better understanding of how some special kinds of behavior—for instance, altruistic behavior—have arisen in the course of evolution by means of natural selection. The last chapter of Becker's *The Economic Approach to Human Behavior* expatiates on this issue, one of the most discussed conundrums of sociobiology, using mathematical models of economic optimization (or utility maximization) to explain not only the origins of altruistic behavior but also the reason why the economic concept of utility maximization can shed brighter light on the biological concept of survival value:

> I have argued that both economics and sociobiology would gain from combining the analytical techniques of economists with the techniques in population genetics, entomology, and other biological foundations of sociobiology. The preferences taken as given by economists and vaguely attributed to "human nature" or something similar—the emphasis on self-interest, altruism toward kin, social distinction, and other enduring aspects of preferences—may be largely explained by the selection over time of traits having greater survival value. However, survival value is in turn partly a result of utility maximization in different social and physical environments.[54]

Leaving aside the details, what is worth noting here is that Becker sees the economic approach to human behavior as being based on some assumptions that are justified by evolutionary theory, but at the same time, he sees it as a way to clarify a notion, survival value, that lies at the bottom of evolutionary theory, as though the two genres of scientific inquiry—economics and biology—shared the same idiom, or at least the same grammar.

Furthermore, as Becker makes plain in the introduction to his book on the economic approach, he considers his theory a framework for understanding *every* human behavior. For him, the science of economics represents a bridge between biological thinking and the social sciences in general. Therefore, economics plays a privileged role among all the sciences of behavior, which include psychology, sociology, anthropology, legal theory, and political science. According to Becker, in fact, it is economic analysis—rather than psychological analysis—that helps us to grasp the true mechanisms of human behavior. This is perhaps the reason why he did not want to associate with Skinner. And yet, interestingly, some key points made by the two scientists are the same: human behavior has a primary biological meaning; human behavior regularly follows the law of survival value maximization (or, in economic terms, of utility maximization); and human behavior does not entail the existence of conscious minds that are aware of the consequences of their own behavior:

> The economic approach does not assume that decisions units are necessarily conscious of their efforts to maximize or can verbalize or otherwise describe in an informative way reasons for the systematic patterns in their behavior. Thus it is consistent with the emphasis on the subconscious in modern psychology and with the distinction between manifest and latent functions in sociology. . . .
>
> All human behavior can be viewed as involving participants who maximize their utility from a stable set of preferences and accumulate an optimal amount of information and other inputs in a variety of markets.
>
> If this argument is correct, the economic approach provides a unified framework for understanding behavior that has long been sought by and eluded Bentham, Comte, Marx and others.[55]

Becker's skepticism about people's capacity to be aware of the process of behavior optimization, or utility maximization, which goes on regardless of what people think they are doing, gives expression to a belief relatively common among today's social scientists: the "autonomous man," as Skinner put it, cannot be trusted and, in all likelihood, does not even exist.

The problem, however, is that Western political systems and institutions are grounded in the idea that the autonomous man does exist. Modern democracy entirely hinges on this concept, which is increasingly questioned by modern science. As a result, there is a growing divergence between the political-institutional foundations of Western societies and today's social research, of which the new biopolitical regime takes advantage.

6.7 Depersonalization and Disembodiment

Skinner's view, compared to Becker's approach, may seem vitiated by a disproportionate emphasis on the futility of the "autonomous man." Yet, in many respects, his theory is more instructive, in that it gives us more hints about the way that human beings and society are being reconfigured under the new biopolitical regime.

One thing that Skinner shows better than anyone else is how the government of living beings is morphing into a government of modular living beings, or of discrete behaviors. This passage marks the end of the age of biopolitical normalization and the beginning of a new era of ethopolitical optimization. In this situation, what is doomed to vanish is the individual as such. Whether one thinks of individual persons as moral-rational agents or of individual organisms as integrated wholes, they all lose their ontological integrity. Depersonalization and disembodiment become the distinctive traits of the coming population.

Depersonalization of human beings is fostered, in particular, through the adoption of a behavioral approach in those areas of research, from psychology to economics, which most contribute to the biopolitical government of society. The result is that, first, individuals are decomposed into behavioral units to be studied and managed on a populational scale; and second, under such circumstances, people learn to look at themselves and the others in this fashion. Thus, each of them becomes "a life," the singular of the collective noun "population."[56] "A life" is that which behaves, period. Within the conceptual framework that underpins the new ethopolitical regime, indeed, Life manifests itself in behaviors, the elementary modules of which living beings are composed. This means that living beings, including humans, are to be seen as behavioral repertoires, or aggregates of behavioral traits, which become meaningful only when they are taken en masse and combined into "populations of entities" that can be thought of as alive—typically, those that evolve by natural selection. Here, the individual is not truly alive, because populations of behaviors are all that exists. No wonder, then, that the ideas of rational agent, moral person, and the like become obsolete. "A life" denotes the ontological dissolution of individuals into populations.

Such a dissolution leads not only to depersonalization but also to disembodiment. If "body" means the whole of an organism viewed as an integrated living system, then the ethopolitics of our time heads toward an increasing disembodiment of everything that is alive, in accord with the neo-Darwinian doctrine that deprives the organism of any inherent cohesion and reduces it to a collection of adaptive traits to be analyzed separately. As already noted, adaptive traits too can be described in terms of behavioral traits; and the latter, just like the former, can be abstracted away from any particular embodiment. Behaviors, like adaptations, have a life of their own, a modular life unrelated to any organic *Bauplan*, a life that is visible only at the level of populations. This explains why, for Skinner and neo-Darwinian biologists, traits—whether adaptations or behaviors—can be modified one by one, as though they were interchangeable units. At the same time, this explains why, for Skinner and many Artificial Life researchers, Life can be breathed into "machines made by living things," provided that the latter's behaviors are programmed to ensure survival value maximization. Here, again, "a life" does not coincide with any natural organism. Rather, "a life" denotes all that remains of an embodied organism after its ontological disintegration into a population of behavioral traits.

The fact that "a life" may be artificial attests that depersonalization and disembodiment end in a paradoxical denaturalization of Life, which may seem at variance with the dogmatic naturalism of Life on which I have focused in part I of this book. The contradiction is only apparent, however, because such a dogmatic naturalism of

Life entails that Life itself can be abstracted away from any particular medium and, therefore, transcends the limits of the natural kingdom as we know it. Life does not exhaust its force at any moment in evolution; and it cannot be contained within the natural boundaries of any life form. In this sense, not even the living can be said to be fully alive because none of them can ever reach the plenitude of Life—*das Ding an sich* of our time. As for "a life," whether natural or artificial, it is the blatant manifestation of this precarious condition. In the final analysis, all that is needed for "a life" to be possible is that it affords a process of behavior optimization, or survival value maximization, in which Life itself consists. But this process, in turn, implies that "a life" will misbehave on a regular basis, for only misbehavior opens the door to behavior optimization. Said in a slightly different way, errors pave the way to the never-ending revelation of Life.

Over the last century, this idea has crossed the mind not only of biologists but also of philosophers, who have stressed several times that "Life itself contains in its very essence the possibility, even the inevitability, of errors."[57] What we witness today is the implementation of this idea in the field of biopolitics, where it takes the form of a three-part statement: all behavior corrects a previous misbehavior; optimization is the best possible definition for this operation; and optimization is brought about and detectable only at the level of populations. Therein lies the difference between the past and the present: ours is a time when the notion that the individual can be the agent of optimization, or take any active role in the evolution of society, wanes.[58]

According to Skinner, the evolution of cultures proves this point beyond all reasonable doubt. Just as life forms evolve by means of natural selection acting on "populations of entities," so do cultures evolve by means of selection by consequences acting on population behaviors. Thus, individuals are placed at the intersection between two lines of evolution that follow the paths traced by two distinct mechanisms of selection and inheritance of traits: genetic on the one side, social on the other. For Skinner and others, social transmission of cultural behaviors is achieved through learning, imitation, and modeling, which also ensure that optimized cultural behaviors will be transmitted to the next generation.[59] The very idea of social transmission seems to imply that something is transmitted from one individual to the other. And Skinner, in fact, cannot avoid using the word "individual" in his works.[60] But the point is that, from a radical behavioristic perspective, individuals become immaterial entities that have no ontological essence of their own, for they find themselves at the crossroads between two lines of selection by consequences, genetic and cultural, from which they arise in the form of a combinatorial product that will be subject, then, to further (operant) conditioning. Seen in this light, individuals look like ephemeral patchworks that do not possess any ontological autonomy, nor any independence from each other. A kind of ontological collectivism pervades Skinner's view of society.

The concept of "a life," as I intend it, serves precisely to emphasize this ontological groundlessness of disembodied and depersonalized individuals. In the end, from Skinner's perspective, "a life" comes to the fore and stands out from the rest of society only when one's behavior is not in tune with the evolution of species and culture that finds expression in a certain society and that always tends to optimize population behaviors by maximizing their survival value. It is right at that moment that "a life" in the singular gains visibility while continuing to lack any ontological autonomy. When the individual's behavior appears nonoptimal (i.e., maladaptive), and hence disconnected from the rest of society, the very individuality of the individual comes into light. But at that point, according to Skinner, it is time for the behavioral therapist to intervene and apply the technique of operant conditioning to optimize behavior and align "a life" with the whole of society. "A life" is, indeed, the singular of a collective noun that does not tolerate any singularity.

Things become more complicated, Skinner acknowledges, when it is the society as a whole that suffers from behavioral disorders that expose it to the danger of extinction. In this case, as he explains, the behavioral psychologist should be charged with the task of correcting the misbehavior of an entire culture. The technology of behavior control should be used to optimize population behavior and align society with the process of Life. Is that feasible? Fifty years ago, when Skinner published his manifesto against the "autonomous man," this project seemed rather abstruse to most readers. Today, particularly after the COVID-19 pandemic and the introduction of a plethora of behavioral regulations in Western societies that call into question the principles of freedom and dignity, his dream is perhaps becoming reality. Or at least, it now seems less unrealistic that the society of the future will somehow resemble his vision and bring to completion the sacrifice of the Western individual on the altar of Life. Friedrich Nietzsche would have described this transition as a transvaluation of all values, he who considered "*life* in this world to be the sole locus of value, and its preservation, flourishing, and above all its enhancement to be ultimately decisive for determinations of value."[61] Skinner put it more soberly: "A scientific conception of human behavior dictates one practice, a philosophy of personal freedom another."[62]

6.8 Mythic Metonymy

Before concluding, it is worth taking some time to highlight how, contrary to all expectations, the scientific conception exalted by Skinner and the transformation of society that we witness today contribute to reviving obsolete styles of thought and archaic, cryptomythological beliefs. Elsewhere, I have argued that a new grammar of power and knowledge is becoming predominant in the Western world, altering the language of science and politics, and thereby modifying the way in which people are inclined to see themselves and the whole of society.[63] This grammar is characterized by an unconventional logic, a bi-logic, which is typical of certain psychopathological disorders. It makes it impossible to distinguish between the subset of a given set and the set itself, or between the part and the whole, which both dissolve into the same all-encompassing continuum. In this situation, it appears, for example, that no clear-cut distinction can be made between the individual and the society, for they merge into the continuum of the human population, made of depersonalized, disembodied behaviors that are taken for the desubstantialized substance of both individuals and societies.

This desubstantialization has heavy consequences, to be sure. With regard to individuals, each of them morphs into "a life," the singular of the human population, a collective entity that, being framed as a continuum of population behaviors, cannot make room for any individuality proper. As a result, one has the feeling that individuals are insignificant, that they are like zombies, and that "a life" is always less than a life worth living. We may call this a feeling of death-in-life. Illich remembered Erich Fromm emphasizing the paradox of a culture that worships Life but understands it only in terms of survival, thus betraying a necrophiliac obsession with death-in-life:

> At a time when I was still completely unprepared to reflect on the appearance of *life* in public discourse, I suppose it was in 1969, perhaps in 1970, [Fromm] came back to Mexico from a short visit to New York and told me that the word which had most frightened him was the term *survival*—the *survival* of the human race, the *survival* of certain parts of the human race, the *survival* of cultures. He compared this usage with the ancient Egyptians' efforts to build pyramids where their king's soul would survive, and called it a terribly necrophilic term.[64]

Pyramids are eclipsed by secular monuments nowadays, but the paradox of a culture obsessed with survival remains, and with it, the impression that death-in-life is the hallmark of our age. The idea that "a life" never meets the requirements of Life, though it continues to be the sole manifestation of the invaluable value of Life, can take various forms, nourishing the wildest fantasies as well as the darkest nightmares. One of them brings us back to the issue of Artificial Life: it is the transhumanist dream of a new epiphany of Life, half-natural and half-artificial, that one day may allow human beings to reach "a life" no longer ruined by the toxic feeling of being half alive and half dead. The problem with this kind of lucubrations, however, is that they intensify rather than assuage this feeling.

It is well known that the neo-Darwinian biologist Julian Huxley was a forerunner of transhumanism, and we can take this as a sign that the idea of an *artificialization* of Life is closely related to the cryptometaphysical *abstraction* of Life from any particular medium on which the neo-Darwinian research programme is based.[65] Unsurprisingly, this is also a sign that transhumanist promises, like previous speculations about Life, are poisoned with threats. According to transhumanists, granting that Life "is just bytes and bytes and bytes of digital information," as some exponents of neo-Darwinism claim, it can be expected that Life will bid farewell to human beings at some point in the distant future and find an altogether different medium in which to evolve.[66] As a leading advocate of the robotic revolution has prognosticated, we are getting closer and closer to the moment when humanity will finally disappear from the face of the earth as a result of a divorce between Life and natural organisms:

> We are very near to the time when virtually no essential human function, physical or mental, will lack an artificial counterpart. The embodiment of this convergence of cultural developments will be the intelligent robot, a machine that can think and act as a human, however inhuman it may be in physical or mental detail. Such machines could carry on our cultural evolution, including their own

construction and increasingly rapid self-improvement, without us, and without the genes that built us. When that happens, our DNA will find itself out of a job, having lost the evolutionary race to a new kind of competition.[67]

The dual characterization of Life as a natural-and-supranatural thing-in-itself, which is intrinsic to living beings but also transcends them, is by no means an exclusive feature of transhumanism. Importantly, moreover, this dual characterization is not just a matter of academic speculation. On the contrary, transhumanism brings into the open a tacit, yet widespread attitude toward Life that fuels a feeling of anxiety among scholars as well as laypeople, partly explaining their remissive acceptance of new forms of social control. Today, Life is perceived inadvertently as a deity that towers over the natural world, commanding an endless optimization of behavior and making a ruthless selection among living beings. Those who transgress its commands will automatically cease to exist or, worse still, discover all of a sudden that they are already half dead.

Some thirty years ago, Illich was one of the first to realize that a new biopolitical regime was being established on the basis of this common belief in a new "pseudo-god." He was vocal in his opposition to it, denouncing the false certainties on which the new "managing clergy" leans while endeavoring to gain complete control over society. According to Illich, a renewed effort on the part of critical theorists is needed to fight this unprecedented configuration of biopolitical power:

The epoch in which we live is abstract and disembodied. The certainties on which it rests are largely senseless. Their worldwide acceptance gives them a semblance of independence from history and culture. What I want to call *epistemological askesis* opens the path toward renouncing those axiomatic certainties on which the contemporary worldview rests.[68]

No doubt, here and in his later interviews, Illich made a good point: critical thinkers have the duty to debunk the irrational certainties from which the modern idolatry of Life draws its strength. That said, his plea for "epistemological askesis" is not very convincing for the simple reason that "the contemporary worldview," as he calls it, is grounded in a confusion between the epistemological and the axiological regimes of enunciation. Life is not only an "epistemological indicator," as Foucault says, but also an axiological indicator. Thus, things are more complicated than Illich thought. Epistemological askesis would be a viable option if it did not entail the repeal of one's overall orientation in life, and if there were valid axiological alternatives at one's disposal. For Illich, a fervent Christian who had faith in the God "who took on flesh and who redeemed us," such an alternative was surely available. But this is not the case with most people today, for whom renouncing the axiomatic certainties of our time may not be that simple.

There is also another reason to be skeptical about epistemological askesis—one that takes us back to the syntax, or grammar, of the certainties that Illich deems senseless. In part I of this book, emphasis was already placed on the current revival of archaic modes of thinking such as paleological reasoning and anthropomorphism: the former consists in reading natural phenomena in terms of teleological processes; the latter, in the personification of natural forces. But there is yet another aspect of primitive thought that deserves close attention. Ernst Cassirer described it as follows:

> In mythico-linguistic thought . . . we find in operation a law which might actually be called the law of the levelling and extinction of specific differences. Every part of a whole is the whole itself; every specimen is equivalent to the entire species. The part does not merely represent the whole, or the specimen its class; they are identical with the totality to which they belong; not merely as mediating aids to reflective thought, but as genuine presences which actually contain the power, significance and efficacy of the whole. Here one is reminded forcefully of the principle which might be called the basic principle of verbal as well as mythic "metaphor"—the principle of *pars pro toto*. It is a familiar fact

that all mythic thinking is governed and permeated by this principle. Whoever has brought any part of a whole into his power has thereby acquired power, in the magical sense, over the whole itself. . . .

If, therefore, ancient rhetoric names as one of the principal types of metaphor the substitution of a part for the whole, or vice versa, it is easy enough to see how this sort of metaphor arises directly out of the essential attitude of the mythic mind. But it is equally clear that for mythic thinking there is much more in metaphor than a bare "substitution," a mere rhetorical figure of speech; that what seems to our subsequent reflection as a sheer transcription is mythically conceived as a genuine and direct identification.[69]

As Cassirer explains in this passage and elsewhere, mythic metaphor is essential to mythic thinking. But a mythic metaphor is not a metaphor as we ordinarily understand it, because the mythic mind takes this figure of speech *literally*. This means that the *pars pro toto* relation is taken for a *pars qua totum* identity: a part and the whole become the same thing, while remaining distinct entities. I myself had examined this and other literalized figures of speech in my essay on political grammars. Therein, I had defined the *pars qua totum* figure as a delusional metonymy, for I consider the *pars pro toto* to be a specimen of metonymy rather than a metaphor.[70] For this reason, I will speak of "mythic metonymy" from now on.

The role played by mythic metonymy in the "contemporary worldview" should not be overlooked, however illogical this concept may seem from the point of view of educated people who have been taught to follow the principles of rational thinking. It is indeed such a paradoxical figure of speech that connects "a life" with Life as such, leading to a cultural regression that combines technological progress with mythicolinguistic thought. The illogical nature of mythic metonymy, and the reason why it is so difficult to discern how it works (and *that* it works), lie in the fact that it lends itself to two opposite interpretations, which both are valid at the very same time, as attested by the example that concerns us directly.

On the one hand, "a life" participates in the evolution of Life, in that it always struggles for survival. In this sense, "a life" is to be construed as a partial, local manifestation of Life in general. Life cannot be contained within the boundaries of any single living being or life form, and yet it finds expression in each of them. From this perspective, "a life" cannot be identified with Life, even though we use the same word for both entities. Rather, "a life" carries a figurative meaning: literally, it denotes each and every living being; figuratively, it refers to Life as such, to the very Life of "a life." In this manner, a figurative *pars pro toto* relation between "a life" and Life is established.

On the other hand, "a life" is also a manifestation of Life in its integrality because the very existence of Life, of a universal force that extends its jurisdiction into the whole of the natural world, would be threatened by any single living being that proved capable of opposing its power. If just one living being could disobey Life's commandments, Life understood as a ubiquitous force that rules over everything that is alive would cease to exist. From this perspective, therefore, it is the totality of Life that flows into "a life." Since Life admits no exception, "a life" identifies with Life itself, literally. The nomological and axiological order of Life nature hinges on this *pars qua totum* identity.

The biopolitical regime transforms this piece of mithico-linguistic thought into a powerful tool of government, exploiting the conflation between the nomological and axiological regimes of enunciation. Once "a life" has grown into a mythic metonymy for Life itself, it can be assumed that "a life" not only receives the order of Life but also issues that order and restates it without pause. And this is why "a life" becomes so prone to obedience and submission every time it hears the voice of the new ethopolitical regime: in the latter's voice, "a life" hears its own voice resounding.

Clearly, the new pagan divinity from which the ethopolitical regime of our time draws its legitimacy has no similarity with the Christian God in whom Illich believed, nor with any other divinity that once watched over the world from above, from a distant place beyond the sky. If anything, it bears close resemblance to those mythological deities described by our ancestors in terms of processes or forces (in Latin, *vis*) that were immanent in nature and consubstantial with the animate things on which they exerted their power. As Cicero writes in his *De Natura Deorum* (III 47):

> Natio too must be pronounced a goddess, for when we go on pilgrimage to shrines in the Ardea area, we regularly offer sacrifice to her. She derives her name Natio *a nascentibus*, from those being born, because she protects married women who are in labour. And if Natio is a deity, so are all those abstract figures which you mentioned: Honor, Faith, Mind, Concord, and following these, Hope, Money, and any concept imaginable.[71]

Where is Natio? It is everywhere a child is being born. Likewise, Life is everywhere "a life" arises and gesticulates. The only difference between the ancient goddess and the new one is that the latter has an all-inclusive unity: it embraces *all* aspects of human existence. Nowadays, Life has taken the place not only of Natio but also of Honor, Faith, Mind, and similar deities of the past, thereby turning into a monotheistic mythological deity that closes the door on any other form of religious belief and urges us to revert to paleological thinking and an anthropomorphic conception of Mother Nature. This new deity governs the world in which we live, literally and figuratively. And mythic metonymy is instrumental in making such a world possible.

Short of a straight identification between Life, or Mother Nature, and "a life," or the human being, neither the interpretation of natural phenomena in terms of teleological processes nor the personification of natural forces would be conceivable. A mythic metonymy thus marks the reversion to mythicolinguistic thought, disclosing the truth of our time: the struggle for life is not only a struggle for the survival of "a life" but also (and above all) a struggle for the survival of Life as such. In the end, every time we follow the order of Life, we do nothing but allow Life to come into being; we bring it to light for a while, saving it from the darkness of nothingness. Every time we worship our god, we keep it alive, together with ourselves.

6.9 Exit

In conclusion, one may wonder whether the current "reversion to mythology" can be counteracted, and how.[72] As already remarked, Illich's epistemological askesis cannot be of much help; and the same holds for any critically informed analysis of the "current worldview" or, worse still, any appeal to the rational and moral faculties of the "autonomous man." As a matter of fact, there is nothing that critical thinkers or philosophers can do against the new mythicopolitical power: no theoretical analysis can have practical impact, and no exercise in thought can oppose the intrusive forms of surveillance and behavioral manipulation that find their legitimation in the present-day mythology of Life. In light of all this, however, one can predict with a certain degree of certainty that the new mythicopolitical power will fail to achieve the complete transformation of humans in modular living beings to be handled like living machines.

The reason for this is that the feeling of death-in-life that weighs on "a life" is doomed to become unbearable in the long term. From what has been said so far, it should be clear that "a life" will never be able to comply with the order of Life, because Life is ultimately that which deprives "a life" of any life of its own. Therefore, the shorter the distance between Life and "a life," the larger the distance that separates "a life" from itself. And the stronger the effort of "a life" to comply with the order of Life, the greater its pain and feeling of inadequacy. As a result of such a double-bind situation, our societies will be pushed closer and closer to the brink of collapse with the passing of time, regardless of our being aware of the major spiritual crisis that we are going through. Said in a slightly different way, it seems that there is no need to ask ourselves what is to be done, not only because there is nothing that we can actually do, but also because neurosis—as Sigmund Freud calls mental malaise in general, "in which one aspect of the ego . . . comes into conflict with its other aspect"[73]—will do the job for us.

Neurosis, as Freud explains, is that which draws a line between humans and other animals such as pigeons and rats. "A dissension of this kind may perhaps only occur in human beings, and on that account neurosis may, generally speaking, constitute their prerogative over the animals."[74] If Freud is right in saying that neurosis is a distinctive trait of human beings, then it is likely that neurosis will prevent humans from becoming domesticated animals, or fully automated living machines that follow Life's commands without hesitation and without difficulty, as Skinner theorized.[75] On the contrary, malaise and discontent will grow in the future, owing to the highly pathogenic effects of the new mythicopolitical power. Whether this process will eventually prompt a regeneration of our societies, time will tell. As Johan Huizinga once remarked:

> History can predict nothing except that great changes in human relationships will never come about in the form in which they have been anticipated. We know that the shape of things to come will be different from any that we *can* imagine. In the outcome of any epoch there is always a component which is afterwards understood as the novel, the unexpected, the previously unthinkable. This unknown *may* mean ruin. But as long as expectation can hesitate between ruin and salvation it is our duty to hope.[76]

Conclusion: The Moral of Morals: On Bioethics

For centuries, Life and the body have been inseparable. Today, their relationship to each other gives rise to questions that are at the center of heated debates among bioethicists. Here, I will not attempt to unravel any bioethical problem but rather analyze the metaphysical disjunction that hides in the background and renders some of those problems unsolvable. Among contemporary scholars, there are many who think that metaphysics is dead and philosophers had better not elaborate on metaphysical issues any longer. The fact is, however, that metaphysics infiltrates into scientific thought. In particular, it makes inroads into biological research, thereby affecting the scientific image of the natural world. As will be shown next, this invisible, ghostly return of metaphysics has an impact on the manifest image as well, fashioning our everyday picture of natural phenomena and human beings. In the end, certain bioethical issues prove so difficult to address precisely because of such an unnoticed resurgence of metaphysical assumptions in the field of scientific theorization—namely, because modern biology has decreed that something like Life as such can be abstracted away from any particular medium, including the human body, leaving society unprepared to confront those critical situations in which the distance between Life and the body appears to be maximal. Sometimes it is Life that seems to detach itself from the human body; at other times, it is the human body that seems totally unfit to house Life. In both cases, "a life" arises from Life in the abstract, taking the shape of a thing that wages war on the human body.

C.1 Fetus: Life without Body

The conflict between Life and the human body has been thoroughly investigated by Barbara Duden. As she emphasizes, the idea that the fetus is "a life" that should be protected against abortion, an act often construed as murder, appeared not long ago. Hence, there is the need to reconstruct and analyze in detail the genealogy of what is usually called the "maternal-fetal conflict."

In her essay on pregnancy and the unborn, Duden stresses that the word "life" is not used today in the same way as in the past. In the Christian worldview, for example, life meant the participation in Jesus Christ, the Son of God, who announced to the world: "I am the resurrection and the life. The ones who believe in me will live, even though they die; and whoever lives by believing in me will never die" (John 11:25). As Duden recalls, it is only in the late modern period that a new concept of Life emerges, together with the science—biology—that looks into Life as such, Life abstracted away from any particular living being:

> In most of the New Testament and in two thousand years of ecclesiastical usage, to "have life" means to participate as a believing Christian in the life of Christ, a mode of living which is placed in stark contrast to mere human existence. . . .
>
> Antiquity has no concept or term for substantive life. *Bios* means destiny, curriculum. *Zoe* expresses the vitality and splendor of ensouled beings. *Psyche* can be used to translate the Hebrew word for blood and, occasionally, the soul. Amateur antiquarians who replace "life" in the modern context with one of these words make fools of themselves. . . .

> Aristotle knows no blackness, but he knows black things; he knows no life, but he left us brilliant treatises on beings that are alive in several degrees of intensity—plants, animals, and humans who, in addition to growing and moving, are endowed with speech and especially with laughter. Life as a substantive notion appears a good century after the final demotion of Aristotle as the great science teacher, or about two thousand years after his death. In 1801 Jean-Baptiste Lamarck introduced the term biology into the French language.[1]

In these and other passages of her essay, Duden makes a point that had been made by Michel Foucault some thirty years earlier: before the nineteenth century, Life itself did not exist because all that existed was living beings, not Life as such.[2] To this, Duden adds the remark that Life in the modern sense is an extremely vague and slippery concept. Although the word is employed in various contexts and used as a substantive notion that allegedly refers to something that exists "out there," it is not clear what this something might be. For this reason, Duden avers, the word "life" is likely to invoke, more often than not, a purely emotional response. As she says, Life is a polymorph, "full of connotations but impotent for denotation":[3]

> One only has to be vaguely alert when leafing through glossy mass circulation magazines to recognize how attractive life has become for the advertising industry, how clubs, shoes, towels, toothbrushes, and travel are identified with the quality of "life." . . .
>
> Any criticism is immediately answered by someone who connects life with right or value or sacredness and, by so doing, evokes six million Jews, or sixty million fetuses, or large numbers of Kurds and Cambodians, or even the rain forests, bugs, and grasses.[4]

As a result of this "maniacal brandishing of life," the four-letter word becomes "meaningless and loaded; it can barely be analyzed, yet it is a declaration of war."[5] According to Duden, not even the life sciences are of any help when it comes to clarifying the meaning of Life because "Life appears in molecular biology only in conversation, when, for example, a laboratory director seeks support for a foundation grant from his neighbor at a celebrity dinner. The egg cell becomes a life mainly when DNA is discussed on TV."[6]

To an extent, one can agree with Duden. It is true that the word "life" is rarely employed by molecular biologists when they are at work in the laboratory. It does not follow, however, that the concept of Life is irrelevant to today's biologists, as Duden seems to imply. On the contrary, this concept is crucial for the life sciences, particularly evolutionary biology, as already explained here. And for this reason alone, anyone can witness a scene like the following one, which Duden finds absurd but which is all too common:

> At least during the 1980s, it would be difficult to remember a discussion on the future, religion, democracy, ecology, women, or hygiene without a characteristically absurd scene: an interruption from the podium or the floor by someone stating: "I, as a biologist and a woman . . ." or, "I, as a biologist and a mother . . . must say something of technical importance about life." Occasionally, I have been surprised by the caliber of the scientist who did not blush to enter the fray this way.[7]

What leaves Duden speechless is the fact that many biologists, to say nothing of philosophers of biology, conflate the nomological interpretation of Life with an axiological assessment of Life. As a result, they feel that they are more entitled than anyone else to judge what is good or bad in the moral sense. If this is possible, however, this is not because they have a better knowledge of nature and human beings, but rather because they belong, as all of us do, to a certain culture, modernity, that is marked by a scientific metaphysics that blurs the boundaries between naturality and morality.

Duden, like Ivan Illich before her, never took this metaphysics into serious consideration. Nevertheless, her analyses help us see some of the most paradoxical and annoying consequences stemming from that metaphysics—in particular, the idea that "a life" can exist in the absence of, and sometimes even in conflict with, the human body. This is where the fetus, a human life that has not yet developed into a human body, enters the scenario.

If Life is a modern invention, as Foucault was the first to realize, then the fetus is, from Duden's standpoint, a modern vision that has transformed that invention into a palpable fact. To achieve this result, several steps had to be taken. First, pregnancies had to become the object of medical inspection under pressure from public health authorities that deemed it necessary to control and regulate the demographic growth of the population. Second, new medical techniques of intervention and exploration of the woman's body, now put under strict surveillance, had to be devised and refined with the passing of time. Both steps were taken in Europe during the late eighteenth and the nineteenth centuries, when a new biopolitical conception of the duties and powers of states gradually imposed itself on the continent. As Duden recalls, one of the champions of such a political-medical revolution was the German physician Wilhelm Gottfried Ploucquet (1744–1814):

> Ploucquet sees the physician's task as "discovering pregnancies." According to him, the simplest way to achieve this end would be "the general installation of monthly public baths which should be made obligatory for each unmarried woman who is over fourteen and under forty-eight years of age." Ploucquet deemed this inspection through public bathing "the only real method" for discovering pregnancies. . . .
>
> Ploucquet wants to feel or see what no man has ever before attempted to feel or see, the movement of the fruit. He wants ultimately to bypass society's dependence on the testimony of women. We have become so accustomed to pregnancy tests that we need to stop for a moment to realize Ploucquet's daring. He belongs to the first generation of physicians who in the early nineteenth century palpate a woman's body.[8]

Today, as Duden notes, the public exploration of the woman's body has reached a much higher degree of sophistication. After the introduction of ultrasound technology, in the early 1960s, the unborn entered the field of vision. As a result, not only the woman's body but also the fetus are now exposed to the public gaze, and this contributes to reinforcing the impression that a separate entity, one that is and is not part of the body in which it is contained, inhabits—or, perhaps better said, colonizes—that body. As Duden emphasizes, technological advancements and the subsequent appearance of the "public fetus" have been instrumental in popularizing the belief that the unborn child should be considered, for all intents and purposes, "a life." As she clarifies, "the formation of the fetus is to a large extent the history of its visualization."[9] Yet it seems perfectly reasonable to ask what "a life" without a body can ever be. Worse still, "the historian of the body" can even wonder whether the "abstraction of the fetus" has any meaning at all:[10]

> Ultrasound got into the operating room in the 1970s as a subsidiary instrument for the surgeon and then into the offices of a few gynecologists as a claim to prestige. It became essential for the state-or-the-art prognosis, "user friendly" in a decisive way. The screen was so arranged that the pregnant woman could join her physician in real time to view the inside of her belly. She no longer had to rely on word of mouth or medical judgment to interiorize the emblem from the screen. With her own eyes, she could now pretend to see reality in the cloudy image derived from her insides. . . . The abstraction of the fetus as a "new life" takes on the consistency of a neoplasm.[11]

As Duden goes on to explain, the abstraction of the fetus, however artificial and baseless it may seem to the disenchanted eye of the historian, has nonetheless opened the door to the current sacralization of "a life" without a body—a sacralization sponsored by the Christian Church, both Catholic and Protestant, despite the fact that it has nothing to do with the lore of Christianity. Today, in fact, it is no longer in the name of Jesus Christ and his promise of supranatural life, but rather in the name of a biological understanding of natural Life that the Church urges believers to recognize the fetus as "a unique human life," thereby prioritizing the latter's rights over a woman's rights. Duden is particularly harsh about the 1986 declaration of the German episcopal conference on "The Life of the Unborn Children," from which she quotes the following passage:

> Modern biology has proven beyond any shade of doubt that there is no such thing as the prehuman stage of the embryo in the womb . . . This is not a theological opinion or ideology. It is emphatically a fact that from the moment of conception onward we face the presence of a unique human life. The early stages of the human embryo are in no way early stages toward the human. This insight has been for a long time the common knowledge of biologists.[12]

As Duden notes, the German episcopate refrains from explaining in more detail what this "common knowledge of biologists" is. As a matter of fact, neither the German episcopate nor any other exponents of pro-life movements around the world are used to grounding their arguments in defense of the life of the unborn on a well-defined concept of Life or of the human being. What they learn from biologists is just that Life and embodied living beings are not the same thing, but they would find it immensely difficult to explain the reason for this. That said, the a priori distinction between Life itself and the embodied living on which modern biological theorization is grounded creates the possibility of mapping one disembodied abstraction, Life, onto another, the fetus. The result is "a life" without a body: a visible epiphany of an invisible assumption.

Disembodiment is the most crucial concept in Duden's essay. The fetus is a disembodied entity that grows in the womb of a woman, and it disembodies her too, in that it disowns the latter of her own body. Importantly, disembodiment does not mean disappearance here. As for the pregnant woman, she does not vanish but is reshaped into a depersonalized being deprived of rights, a uterine environment, a hydroponic system that is necessary for the blossoming of "a life" that is not yet a body.[13] Thus, as Duden says, while "in Roman law the human fetus was regarded not as a legal subject but as part of a woman's body,"[14] today the perspective is reversed: a woman is perceived and valued from the point of view of the fetus, with the result that her body is reduced to a biological habitat for "a life," while her legal subjectivity is doomed to retrogress. As for the fetus, it turns into an emblem of Life as such; that is, an emblem of our time.[15] To understand why, one only needs to realize what type of crime abortion is in the eyes of the Church. In the past, the question had often arisen as to whether the unborn had a soul, and at what exact point in time they were ensouled, because only after that moment was the interruption of pregnancy to be considered a form of murder. Specialists charged with the task of finding out the moment when a soul quickens in the unborn were not physicians and biologists, but rather theologians and monks:

> For many centuries, the socially recognized specialists concerned with knowledge about women—today called gynecologists—were mainly monks, men forbidden by vow to enjoy the flesh and by church law to practice medicine. The prime instances of women in labor remained Eve, Mary, and Elizabeth. The theologian looked for the moment of ensoulment—the moment when a new soul is made out of nothing by the Creator but also the awesome instant when original sin is transmitted from Eve's flesh to the new creature.[16]

Today, things have changed. The Church considers abortion not so much the murder of an ensouled being as the impious assassination of "a life." This means, among other things, that the idea of the sacrum and the definition of sacrilege become intertwined with the notion of Life itself and with the sacrosanct epiphany of Life: the fetus.[17] As the German episcopate solemnly declares, from the moment of conception, we are in the presence of "a life," and it is Life that we should protect and value above all else. This is the holy mystery around which the contemporary cult of the fetus revolves.

I call it a *mystery* because, as Duden emphasizes, it would be extremely problematic for anyone to add further precision and illustrate what "a life" is exactly, in what sense it exists on its own, and why something like that should be deemed sacred. The phrase itself is so unclear and loaded that Duden regards it as a pseudo-concept, a "protean linguistic entity" similar to those mystifying catchwords that Uwe Pörksen describes in terms of "amoeba words"—that is, words that "lose all power to denote and acquire an unlimited power to connote" as soon as they start being used inappropriately by the media.[18]

However, contrary to Duden's opinion that the media alone should be held responsible for the corruption of the public discourse that fools people into believing in the existence of "a life," whereas the life sciences should be acquitted from any such charge, it seems safe to say that science too is responsible, at least partly, for the confusion surrounding current debates.[19] Indeed, the abstraction of Life from living beings, which entails the abstraction of "a life" from the body, lies at the core of the life sciences. And therein lies the reason why the fetus constitutes a mystery rather than a mere superstition.

This observation does not invalidate Duden's overall argument. If anything, it ends up strengthening her conclusion that "a life" manifests the sacrum, or the last vestiges of the sacred, in our secular world. In point of fact, even within the framework of ancient civilizations, the sacrum was not something that was totally detached from, or contradicted, the general worldview promoted by savants and shared, to a greater or lesser extent, by the common people. Rather, the sacrum expressed the mystery that surpassed and thereby delimited the knowledge that a certain society had gained about the natural world. The sacrum was, in other words, a liminal space between the known and the unknown, a gray zone that enthralled everybody and required special attention. Likewise, the current belief in the existence of "a life" without a body does not amount to sheer superstition or insane nonsense because it does not contravene the main tenets of today's biology. In the final analysis, it is the a priori distinction between Life and the living, lying at the bottom of modern biology, that paves the way for a parallel distinction between "a life" and the human body, or a fully developed, functioning organism. In this case, too, we find ourselves at the frontiers between the known and the unknown, where scientists and ordinary people meet on an equal footing.

In the last pages of her book, Duden recapitulates and makes one more remark of the utmost importance: the only meaning that Life seems to retain when it is abstracted away from living beings is "survival." This is exactly the same meaning that Life has for mainstream evolutionary biology. And this is, more generally, the basic meaning that "a life" has today. In our world, as Duden points out, to be alive means to survive, no matter whether we are talking about the fetus or the Blue Planet. And not only do the living survive but they are also *called to* survive—they *must* survive, by any means. The order of Life corresponds not only to a natural necessity but also to a moral imperative that commands living beings, from bacteria to humans, to look beyond what they are, beyond their own body, in search and in memory of that pure Life, of that unfathomable creative source, of that body without organs, from which they all originate one after the other. Therein lies their responsibility:

> The growing ecumenical consensus about the sacredness of life is better understood as an aspect of a surreptitious shift in social and medical management concerns about the importance of "survival." Twenty years ago, Erich Fromm pointed out the necrophiliac intensity that emanates from this word. Joy in the aliveness of nature, of the lily, the lark, and the infant, no longer motivates

those who use this terrible word, but rather, fear, calculation, and something Freimut Duve (another German colleague) calls "apocalyptic randiness."

The idol of the fetus has only one competitor at present, and that is the Blue Planet. Just as the sonogram of the fetus stands for one life, so the TV satellite picture of the earth stands for all life. As the pink disk of the zygote appeals for the maintenance of one immune system, so the blue disk of the biosphere appeals for the survival of the entire global system. Both disks act like sacraments for the "real presence" of life, for whose continuation a global "we" is made responsible.[20]

C.2 Wannabe: Body without Life

When it comes to bioethical dilemmas such as abortion, the first question to ask is not how to solve them, but rather why they appear in the first place and why we get stuck with them. Indeed, it is in the dilemma itself, not in the solution, that society can see itself mirrored. This also holds true for the second dilemma discussed here, which became a subject of public debate some forty years ago.

June 21, 1984, was a day of joy for Robert Vickers. When he recalls it, he says that it could have been a tragedy had it not been for the strength of those around him. Fortunately, it was a success: this was the day for which he had been waiting for thirty years, the day when he finally achieved his dream. That day, he was born into a new body: a body without a leg.

The countdown had begun the day before, with shame: he had lied to his wife and his employer concerning his intentions. He stocked up everything he needed to reach his goal. In the late afternoon, he was already satisfied, being sure that his left leg was frozen to the knee. He called his beloved Elaine and told her to pick him up at his mother's place. He did not have a clue about his wife's reaction to what could have seemed an intolerable act of self-mutilation.

Her first reaction was one of seraphic composure: she calmly handled the situation, carried him home, put their five-year-old daughter to bed, and called the ambulance to take him to the hospital. Being a "walking wounded," he was not noticed upon arriving in the waiting room. Medical staff treated him professionally for the whole period of hospitalization, which lasted two weeks. The fact that he had inflicted this mutilation upon himself was never considered an issue. At one point in the evening, Elaine went back home, to their daughter. After some preliminary discussions, Robert was taken to the operating room. A few hours later, he was brought back, but his leg was still there. His initial disheartening was lifted upon hearing that doctors had realized that it was impossible to save the knee. Robert's leg had to be amputated midthigh. There was no question of attempting plastic surgery.

When he woke up after the intervention, Robert found himself in paradise: the leg that he had so much hated since he was a child was no longer there. He did not know where it was, and he did not care. He was focused on his future life. With this surgical operation, he had removed his depression and sadness, suicidal thoughts, and self-hatred. This moment marked for him the beginning of a new life—a life that he had always desired. During hospitalization, his behavior was irreproachable: he did everything that the doctors and nurses asked him to do, being perfectly aware that they were helping him to achieve his objective and attain what might have seemed a disabled life to others. From his point of view, however, he had freed himself from disability. As soon as he could, he took his crutches and left the hospital.

Today, looking back, he knows that his decision might seem egoist to most people, but he remains convinced that he saved his own life. He never asked those around him what they thought about his choice. The only thing that matters is that his life has finally become livable, almost marvelous. After amputation, it is as if he could do things like climbing trees. Even though the whole story may appear outrageous to those who have lost limbs because of an accident, amputation was for him the fulfillment of a long-standing dream. No regrets.[21]

According to some psychiatrists, people who become disabled by choice suffer from body integrity identity disorder (BIID), a mental disease unknown until not long ago, to which health professionals, bioethicists, and legal scholars have drawn attention over recent years. The problem is not only how to describe this mental disorder in theoretical and/or operational terms, but also whether people who have a persistent desire for amputation should always be treated as mentally ill.

The concept of BIID stems from the concept of gender identity disorder (GID), owing to the apparent similarities between the two conditions: in the latter case, it is a matter of changing one's own body by modifying its sex characteristics (GID); in the former, it involves maiming one's body in one way or another (BIID). In one case, it has become customary to talk about transsexualism; in the other, the symmetrical concept of transableism has been proposed.

The amputation of healthy limbs is not the only goal that BIID people may want to achieve. Some of them dream of becoming deaf; others, blind; yet others, paraplegic. Considering such a diversity, the concept of BIID is now used to refer to individuals with an intense desire to acquire any kind of physical disability.[22] It is not easy to estimate how widespread transableism is nowadays, nor is it clear whether the same type of disorder, or desire, can also be found in non-Western societies.[23] The research is ongoing.

Michael First was among the earliest psychiatrists who examined in some detail this clinical condition. In his pathbreaking article of 2004, he emphasized that most people affected by transableism are not prompted to harm themselves by atypical sexual interests (paraphilia), but rather by the feeling of a mismatch between one's anatomy and one's sense of identity. Based on his research, he discarded the hypothesis that these individuals can be labeled delusional following current diagnostic criteria. As a matter of fact, they often realize how abnormal and shocking their desire may appear to others; moreover, only in very few cases are they motivated by the desire to find themselves in need of assistance for the rest of their lives. Among the subjects interviewed, none reported noteworthy episodes of mania or hallucinations: "Most had no significant psychiatric symptoms (apart from their preoccupation with amputation) or drug or alcohol problems. The remaining subjects had mild symptoms such as depression and anxiety."[24] Many had been in psychotherapy at some time in their lives, but almost half never told their therapists about their desire for amputation, fearing that the therapist would consider such a desire evidence of severe mental illness. A total of 40 percent of the interviewees had taken psychotropic medication, usually for depression, which had no appreciable effect on their condition. In his paper, First also provided a list of the desired locations for amputation: right or left limb, above or below the knee, one or both legs, and so forth. Among the interviewees, six "had a major limb amputation at their desired site and reported that following the amputation they no longer had any desire for an amputation and that they felt better than they have ever felt."[25]

The main conclusions that First drew at the end of his pioneering work are as follows: first, since no established definition in the DSM-IV allowed specialists to classify and understand the behavior of these subjects in terms of mental disorder, it was advisable to introduce a new nosological concept (namely, BIID); second, since psychotherapy and drugs prove ineffective in the vast majority of such cases, the question should be posed as to whether surgical interventions are to be ruled out categorically. The very last words of First's article invite reflection on this topic: "A more provocative question is whether, as a last resort, surgery should be considered as a potential treatment for this disorder."[26]

It is worth recalling that First was not the first to take surgical treatment into serious consideration. Some years earlier, the problem had come into the public eye after a surgeon in Scotland had decided to amputate the healthy legs of two men. The *British Medical Journal* reported the news, highlighting the unprecedented situation with which medical institutions had been confronted:

> A surgeon in Scotland amputated the legs of two psychologically disturbed men who had nothing physically wrong with them but felt a "desperate" need to be amputees, it emerged this week.
>
> Both men, one from England and one from Germany, had a rare type of body dysmorphic disorder known as apotemnophilia, in which patients are convinced from childhood that they will be normal only once a limb has been removed. The obsession is always with the removal of a specific limb, and each patient had a leg amputated above the knee.
>
> The operations were carried out in September 1997 and April 1999 at an NHS hospital, Falkirk and District Royal Infirmary, by consultant surgeon Robert Smith. Both men had been turned away by other doctors. . . .
>
> He said that the patients' lives had been transformed by losing a limb and they were delighted with their new state. Both had had artificial limbs fitted, though they did not always wear them.
>
> Earlier, he told a press conference at the hospital: "At the end of the day I have no doubt that what I was doing was the correct thing for those patients."
>
> The trust's chairman, Ian Mullen, said such operations were not ruled out for the future, but a strict procedure would have to be followed.[27]

Shortly after the news from Scotland came out, a heated debate broke out.[28] Some scholars subscribed to the idea that surgery should be considered a therapeutic option, whereas others argued vehemently against it. Then, in 2013, DSM-5 was released. Contrary to First's advice, it was decided not to include the diagnosis of BIID. The editors, however, took another decision of no less importance: they excised the diagnosis of GID and replaced it with the new label of "gender dysphoria," thereby implying that discontent with one's gender was not a disease in itself, despite its placement in a psychiatric classification of mental disorders.[29] Some years later, the diagnosis of BIID finally appeared in the eleventh revision of the International Classification of Diseases (ICD-11) but with a different name, "body integrity dysphoria." The implication was the same as with gender dysphoria: in both cases, psychiatrists found it impossible to draw neat boundaries between mental disorder and mental health, normal and abnormal.

It is worth noting at this point that, besides some similarities, there are huge dissimilarities between the two types of dysphoria. Among them, the most important is that people affected by transableism, unlike transsexuals, cherish a dream of pure loss. Roughly put, transsexuals feel it necessary to modify their gendered bodies and seek medical assistance for the purpose of transitioning to the sex or gender with which they identify. For this reason, they request interventions of bodily integration, from hormone treatments to plastic surgery, and not only amputation (emasculation). Transsexual individuals desire to transform their bodies rather than merely compromise its functioning. For people suffering from BIID, on the other hand, the problem is not the nonidentity between gender and the body, but rather the perceived contrast between Life and the body.

In other words, "wannabes"—as individuals who have a persistent desire for amputation sometimes refer to themselves—do not aim to correct and ameliorate their bodies; they do not want to *add* anything to their bodies and change them to bodies with different, more suitable characteristics. In the end, they just want to disable their own bodies. For them, the point is that Life and the body do not get along, and the only way in which Life can be preserved is by distancing it from the body. This is why, from their point of view, the dream of pure loss turns out to be a dream of pure benefit: the dream of "a life" finally disembodied, the dream of a survival of Life itself, freed and abstracted away from any particular, disposable medium.

This interpretation may seem hyperbolic or sound far too theoretical; yet it is rather close to what transabled people say about their condition. Before amputation, wannabes feel hopeless, as though they are encumbered by too-heavy bodies that impede them from living, literally—hence the high risk of self-harm or suicide. After amputation, transabled people give the impression that they have experienced a significant and lasting increase in well-being. Some declare themselves to be less desperate; others to be utterly happy.[30] In the aftermath of mutilation, moreover, most of them stop ruminating over self-injuries. It has been observed, perhaps correctly, that evidence is merely anecdotal because the number of patients who have secured surgical treatment is too small.[31] But the fact remains that on the basis of the scant evidence available, many specialists are now inclined to believe that the option of elective amputation in a protected environment (e.g., a hospital or medical facility) should not be excluded from the outset.[32] In light of this new awareness, Life as such begins to enjoy a very privileged status compared to the human body, not only for patients but also for health professionals. Gradually, the sacredness of Life takes the place of what was once deemed far more important from both a religious and a medical point of view: the sanctity of the human body, not to be injured by medical doctors (*primum non nocere*) and destined to resurrection after the end of the world.[33]

Interestingly, when one looks at the reported cases of BIID more closely, the very concept of *identity* disorder appears somewhat misleading. We may indeed ask ourselves what wannabes are in search of, or what fills them with such a gratifying sense of relief after self-inflicted mutilation; and the answer is not (true) body identity, as First and others have suggested—not least because self-inflicted injury and consequent disability *disidentify* the human body. Again, the comparison with gender troubles clarifies the point: no positive reconfiguration, or *rehabilitation*, of the body is in sight here. For wannabes, the body must lose, at least partially, its ordinary functionality without gaining any new ability. Furthermore, this weird desire for disability is not inspired in them by an "alter personality," as in the case of dissociative identity disorder, commonly known as "multiple personality disorder," which is the only identity disorder included in DSM-5. Wannabes do not ignore who they are, nor are they under the spell of a supernatural voice or falling prey to a dual personality that divides them from themselves. In general, their conduct seems not divorced from reality, and psychiatrists cannot diagnose wannabes as psychotic patients on the grounds of the currently available definitions of psychotic disorders. Thus, the question comes up again: If wannabes are not mentally ill, where does their desire to *disfigure* their own body come from?

In the final analysis, there are only two ways to address this problem: either we decide that the time has come to revise the established definitions of psychosis, delusion, and the like; or, after discarding all extant diagnostic criteria, one after the other (paraphilia, factitious disorder, dysmorphic disorder, psychotic disorder, and so on), we take the wannabes' words for the diagnosis itself: "identity disorder." A trouble with identity is in fact the reason that they generally provide for their own desire for amputation. First and others conclude that this is also the real cause of such a desire.[34] The weakness of this conclusion is, however, quite evident: it leaves unexplained what wannabes *gain* in terms of identity from amputation or self-injury. Given that they want to experience nothing but a *loss*, it is hard to see how such a partial disintegration and disidentification of their own body might strengthen—or corroborate, or authenticate—their body identity. No wonder, then, that research has slowly shifted from the field of psychiatric inquiry to the field of neurological investigation over the last few years. Today, the idea that a neurological syndrome hides behind this mental disorder gains ground among researchers, with the result that the concept of body integrity identity disorder is often replaced with that of "xenomelia." Thus, a neuroscientific taxonomy substitutes for a psychiatric classification.[35]

Leaving aside the astonishing silence that surrounds Sigmund Freud's theory of the unconscious mind, a theory that might be of some help here and account for the likely discrepancy between the motives and causes of the wannabes' desire for self-inflicted injury, recent debates about BIID are striking for another reason: not only do they give evidence of the limits of psychiatric classification but they also throw light on a tacit assumption that is shared nowadays by both wannabes and society at large, including health professionals. This common assumption—which makes it so difficult for specialists to put enough distance between themselves and these patients so as to look at the latter's condition with a sufficient degree of clarity—says that "a life" should not be conflated with the human body. Based on this problematic assumption, none can see right along the insanity of the wannabes' desire. This does not mean that First, for example, does not see it, but it is a fact that drawing a line between the healthy and the pathological becomes particularly arduous at the moment when "a life" and the human body are understood as being relatively independent from one another. With wannabes, specialists may even feel that they are not facing a disease, properly speaking, but rather an unusual yet legitimate request, a rare yet rightful expression of the human pursuit of happiness.

In this case, the expression is no doubt eccentric but not completely inadmissible. It combines two metaphysical principles, the autonomy of the human will and the autonomy of Life as such, into a single riddle. On the one hand, wannabes proclaim the autonomy of Life from any particular medium by fighting against their own bodies. On the other, they proclaim the autonomy of the human will from any particular end—the true achievement of moral autonomy, according to Immanuel Kant—by cherishing a dream of pure loss. Wannabes do not want anything *more* for themselves; in fact, they want *less*, and ultimately they aim at making nothing appear instead of something.

In this manner, wannabes show us where the moral autonomy of the human will lead when it is left to itself and somehow absolutized. When moral autonomy is taken to extremes, as G. W. F. Hegel explains, humans want nothing but their own will, which thereby reduces itself to the will of nothing, a will devoid of content and totally detached from reality, a will that withdraws into itself and turns its back on the world. With wannabes, this will turn its back on its own body as well, and thus gives voice to "the void" and the drive toward dissolution, which, Hegel maintains, lie at the heart of modern civilization—exactly the opposite of a search for identity:

> The will contains the element of *pure indeterminacy*, or of the I's pure reflection into itself, in which every limitation, every content, whether present immediately through nature, through needs, desires, and drives, or given and determined in some other way, is dissolved; this is the limitless infinity of *absolute abstraction* or *universality*, the pure thinking of oneself. . . .
>
> Only *one aspect* of the will is defined here—namely this *absolute possibility* of *abstracting* from every determination in which I find myself or which I have posited in myself, the flight from every content as a limitation. . . .
>
> This is the freedom of the void, which is raised to the status of an actual shape and passion.[36]

If transabled people represent a challenge for modern society, this is not only because they offer themselves at the altar of Life, morphing into "a life," but also because they give tangible expression to "the freedom of the void," which pushes Western civilization forward and yet condemns it to dissolution, as Hegel warned us of long ago. Perhaps this explains why scholars find it so hard to solve this bioethical problem.

The reason is autonomy, the cornerstone of modern civilization. More precisely, the reason is that wannabes cannot be denied moral and legal autonomy unless they are diagnosed as mentally ill and judged to be incapable of understanding the nature and effects of their own decisions. Given that such a diagnosis cannot be made on the grounds of the current classification of mental disorders, the logical conclusion is that wannabes should preserve the right to self-determination even though this means self-mutilation—or so many believe. Moreover, if medical institutions could not give them the possibility of being mutilated in a safe environment, in all likelihood they would attempt to achieve the same objective in a far less responsible manner. Based on these arguments, which are raised time and again in the literature, surgical treatments appear as more than an option; they become a duty for health professionals.

In the Anglophone world, the argument for moral and legal autonomy is often traced to John Stuart Mill, the champion of negative liberty and freedom from external coercion. As Mill famously stated, "Over himself, over his own body and mind, the individual is sovereign."[37] More recently, the legal theorist Joel Feinberg restated the concept and further developed Mill's antipaternalistic approach, emphasizing that "legal paternalism is an unacceptable policy because in attempting to impose upon a man an external conception of his own good, it is very likely to be self-defeating."[38] In short, if adults are treated as children, they will come in time to be like children, losing the power of rational judgment. Nobody can teach someone else what is good and what is bad. This also holds for self-inflicted harm, which is not to be counted as harm at all, so long as it is the result of a "chosen action":

> Chosen actions are those that are decided upon by *deliberation*, and that is a process that requires time, information, a clear head, and highly developed rational faculties. When I use such phrases then as "voluntary act," "free and genuine

> consent," and so on, I refer to acts that are more than "voluntary" in the Aristotelian sense, acts that Aristotle himself would call "deliberately chosen." Such acts not only have their origin "in the agent," they also represent him faithfully in some important way: they express his settled values and preferences. In the fullest sense, therefore, they are actions for which he can take responsibility.[39]

As should be clear from all this, one cannot see wannabes as irrational subjects that lack the capacity of "voluntary acts," starting from Feinberg's characterization of "chosen actions." And, based on his definitions, a case for elective amputation of healthy limbs can be made. As several psychiatrists, bioethicists, and legal theorists have stated over recent years, the notion of moral autonomy speaks in favor of transableism.[40] Minor caveats or disagreements among authors are irrelevant to the bulk of the argument.

There remains only the question of why so many psychiatrists and bioethicists, sometimes the same ones who fight tooth and nail to preserve the wannabes' autonomy, feel somewhat uncomfortable when a decision has to be made and endeavor to figure out how to handle the problem in a less harmful way, as though the concept of moral autonomy reached its limits here and left a sour taste in everybody's mouth.[41] In all likelihood, the main reason for perplexity and unease is to be found in the uncompromising negativism of the wannabes' desire. In this case, the disjunction between Life and the body spurs the desire of "a life" that wages war on its physical support, but such a desire has no further, positive connotation; it creates a vacuum and nothing more. It brings the subject's autonomy to expression, but it also radicalizes the subject's autonomy to such an extent that the individual's desire becomes incomprehensible, if not incommunicable, to others. As a result of this radicalization, wannabes defy our understanding. At the same time, they strike our imagination. In some cases, they even tug at our heartstrings, for we cannot avoid seeing in them a convulsive, yet veridic illustration of the ultimate essence of moral autonomy.

When we look at them from this broader perspective, wannabes suddenly turn into the uncanny emblem of modern civilization—*das Unheimliche*, that which is not only mysterious but also strangely familiar.[42] In some sense, it can even be said that wannabes disclose the "spirit of the time," as Hegel would have it, for they show where modernity ends. They do this by showing the fate of "a life" that rejects its own embodiment for the sake of the abstract autonomy of Life from any particular living and for the sake of the no-less-abstract autonomy of the human will from any particular end. Viewed in this light, transableism cannot be reduced to mere mental disorder. Besides being a clinical enigma, wannabes represent a new "station" in the history of the "spirit" as told by Hegel. From the latter's perspective, they are to be seen as a "shape of spirit"—namely, a culture-specific configuration of "spirit," one that marks the distinctive moment in history when the "shape of consciousness" turns into a "shape of the world." At that moment, Hegel contends, the world manifests its own essence in the self-consciousness of one single individual who is elevated to the rank of a "real spirit." That individual incarnates the truth of the world:

> These shapes distinguish themselves from the preceding as a result of which they are real spirits, genuine actualities, and, instead of being shapes only of consciousness, they are shapes of a world.
>
> The *living ethical* world is spirit in its *truth*.[43]

C.3 Pandemic: On Transmodernity

"Spirit," Hegel avers, is "the individual who is a world."[44] We may wonder, then: What is the world that wannabes incarnate? On the one hand, this is still the world of modernity, the world of Enlightenment, the world of Kantian morality, the world in which moral autonomy reaches its peak by "giving itself the shape of an *impulse*."[45] On the other hand, this is no longer the world of modernity as we have known it for centuries, because wannabes actually close the door on this world: they close it from within, in that they do nothing but draw from the modern notion of moral autonomy the most extreme, paradoxical consequences; but they nonetheless close the door on the world of modernity, in that they show us where modernity *ends*—in the self-annihilating absolutization of moral autonomy. And by drawing the limits of modern civilization from within, they allow us to see the frontiers of a new world that lies beyond modernity. For Hegel, this was the world in which "the *ethical life* of a *people*" finally comes to light, incorporating the world of Kantian morality into a larger framework that does not cancel modernity but rather brings it to completion.[46] For us, two centuries later, the picture is different, and particularly so after the COVID-19 pandemic. What we see rising on the horizon is the moral life of populations, rather than the ethical life of peoples, for it is in the population that "a life," after losing its own body, seems to find a new body, a disembodied body: nobody's body.

In the early months of the coronavirus crisis, when lockdowns and further emergency measures were being adopted in Europe and the US, it rapidly became clear that Western civilization was going through some major changes. On both sides of the Atlantic, well-established constitutional rights that had been deemed unalienable for more than two centuries were suspended all of a sudden. In the following months, the situation worsened.

The German conservative statesman Wolfgang Schäuble was one of the few who sounded the alarm. In an interview released shortly after the breakout of the pandemic in Germany, he stated that "to simply shut everything down for two years would have terrible consequences," adding that subordinating all other concerns to the goal of saving lives was far from being unquestionably correct if only because this sort of "absolutism" contravened the German constitution's right to dignity, that "does not exclude the possibility that we must die."[47]

Among political philosophers and legal scholars, not to speak of politicians, the predominant view was the exact opposite of Schäuble's.[48] Almost everybody was convinced that national governments were fully entitled to take such drastic decisions, detrimental to the people's constitutional rights, on account of the various risks associated with the spreading of the new, relatively unknown virus. The floor was thus open for widely publicized discussions among virologists, epidemiologists, biologists, and medical doctors specializing in infectious diseases. Political decisions had to be taken in accord with the Science of Life to which those specialists gave voice, a science that was promptly transformed into a kind of Hegelian Absolute Knowledge. Nobody was allowed to raise doubts about the newly adopted emergency measures or, sometimes later, the new vaccines that were expected to put an end to the pandemic. The few researchers who dared to object were immediately pushed to the margins, and some of them even lost their jobs.[49] Future historians will certainly engage in furious debate as to whether this frenzy was justified from a medical point of view and fully legitimate from a legal-political point of view. Whatever answer historians give, it is sure that the general reaction to the pandemic prevented people from thinking—at least if we take the word "thinking" in Hannah Arendt's sense:

> In the privacy of his posthumously published notes, Kant wrote: "I do not approve of the rule that if the use of pure reason has proved something, this result should later no longer be doubted as though it were a solid axiom;" and "I do not share the opinion . . . that one should not doubt once one has convinced oneself of something. In pure philosophy this is impossible. *Our mind has a natural aversion against it*" (italics added). From which it follows that the business of thinking is like Penelope's web: it undoes every morning what it has finished the night before.[50]

As Arendt explains, the moment in history when thinking is recognized for what it is—the opposite of certitude—is the modern era, the "era of suspicion," as Friedrich Nietzsche calls it, an era marked by the "universal doubt" that starts haunting both scientific investigations and philosophical meditations.[51] The COVID-19 crisis seems to have put an end to this era. The "universal doubt" has disappeared from the world. However, we may wonder whether it is truly accurate to say that people stopped thinking altogether during the COVID-19 pandemic. It is true that most of them stopped doubting and, therefore, thinking like modern people. But it is equally true that, at some point, they began to think and conduct themselves *in a different manner*, embarking on a journey to an exotic land where a new form of life becomes possible. This land may be called "trans-modernity." Here, modern styles of life and thought are not completely forgotten, but rather reworked and integrated into a conceptual framework which, on one side heads toward the future, and on the other takes us back to the past—a past older than modernity.

The first thing to note is the weakening of the sense of individual identity. Overemphasis on individuality and individualism is a distinctive trait of the modern age compared to the premodern era. But this was no longer so during the pandemic. To begin with, think about the confusion between being infected and becoming ill (or developing a disease related to infection). At the individual level, infection (contagion) does not necessarily lead to illness or compromise one's overall health, which is contingent on many other factors (age, weight, comorbidity, and so on). At the collective level, on the contrary, contagion rate and disease rate are closely intertwined. The higher the contagion rate, the worse the disease looks from an epidemiological point of view; briefly put, contagion rate functions as a detector of the population's health. During the pandemic, the distinction between the two levels became cloudy. The thermometer of the population's health turned into a thermometer of the individual's health, with alternating feelings of anxiety and relief among the public depending on the daily briefing. Perhaps this explains why not only the common folk, but also public authorities and health professionals began to neglect the difference between infection cases and disease cases: whoever tested positive (infection) was considered sick (disease).[52] The infected person thus often regarded herself as an ill person even in the absence of any symptoms.

No doubt the media contributed to exacerbating such irrational feelings, but it would be a mistake to ignore the role played by the new biopolitical technorationality throughout the crisis. In line with the general trend toward population thinking promoted by the biopolitical regime of our time, the epidemiological approach to the pandemic took priority over the medical approach to the disease. From the very beginning, focus was placed on how to stop or contain the epidemic rather than how to treat symptoms and illnesses caused by COVID-19, as though prevention and prophylaxis on a population scale could supplant cure for individuals. The very same logic applied to the issue of rights: it was no longer the citizen who had rights in the first place, but rather the population as a whole.

As a result of this change of perspective, the biopolitical regime found itself in a position to smoothly pursue one of its main goals: optimization. Behavioral optimization was achieved through lockdowns, mask mandates, and further emergency measures. Biological optimization was achieved through mass vaccination some time later. In both cases, it was not at the individual level that optimization could be detected and carried forward. Therefore, it was not on that level that Life was being strengthened. The struggle for survival took place on a population scale, where only the population's body mattered: a disembodied body, "a life," totally submitting to the morals of Life.

The various political, legal, and ethical questions that have arisen in the course of the COVID-19 crisis will continue to be discussed in the years to come. Perhaps attention will also be paid to some metaphysical questions that hide in the background. Consider, for example, the issue of *preparedness*. This concept refers to those actions that are taken as precautionary measures in the face of potential disasters, from hurricanes to epidemics. In the Western world, preparedness has blossomed over the last few years, becoming an integral part of biopolitical techniques. According to some, however, we should do much more to get prepared for the worst-case scenarios with which we might be confronted in the future, for even a tragedy like the COVID-19 pandemic could have been handled much better if we had lent an ear to the advocates of preparedness. As one of them had already warned us in 2006, fourteen years before the spreading of SARS-CoV-2:

> The significance of SARS is *the rate at which it spreads*. It can traverse the globe in a very short period of time, wreaking havoc in places far removed from its origins. There were people in the SARS story of 2003 that the media dubbed "supercarriers," people whose contagiousness significantly spread the virus. Health officials call those people "index cases," but I think *supercarrier* conveys more information.
>
> Modern technology and social networks enable SARS to do greater damage than it otherwise could. What's the worst that it could do? Imagine SARS slamming a pediatric AIDS ward, or a large nursing home. The virus is especially dangerous to the young and the very old because their immune systems aren't strong enough to fight it off. The worst case potential is there, and we ignore it at our peril. Humans cause their own destruction.[53]

With hindsight, one can hardly avoid being struck by this quasi-prophecy. As the author observes, we seem unaware of how many dangers await us in the dark, and we do not do as much as we could to protect ourselves from such mortal threats, nor do we pay "enough attention to the ways we organize society that make us vulnerable to worst cases."[54] The point is, however: What if we become more aware of our vulnerability? Will things get better at that point? Is there anything we can do to prevent future catastrophes? In most cases, unfortunately, we can do very little. Significantly, in the same essay from which this quotation is taken, there is no hint at any concrete way in which we could have limited our vulnerability and reduced the damaging effects of a coming SARS epidemic. This is no surprise since the author is a sociologist, not a public health specialist or emergency manager. But the fact that no such hint is provided, and yet a preventive solution is supposed to be at our disposal, is in itself revelatory of two complementary tendencies in our society.

On the one hand, the power of science is highly overstated. Thus, science, the daughter of doubt, turns into Science, the Mother of Truth. Sometimes scientists are to blame for such a conceited self-celebration. As the leading figure in the global fight against COVID-19 declared at the peak of the crisis, "I am the Science."[55] The failure—or, to put it more gently, the poor results—of many anti-COVID strategies (from lockdowns to mass vaccination) recommended by the Science and promptly implemented by most countries in the Western world proves the opposite. More generally, recent events prove that preparedness should not be confused with shamanism: science, unlike the Science, is far from omnipotent and gives us only a few indications about how to confront nature. If we are inclined to believe in the Science, this is because we wish to protect ourselves not only from nature, but also from our ignorance, which is a source of bewilderment that can easily turn into panic.

On the other hand, the plea for more preparedness coupled with the belief in the infallibility of the Science contributes to making the morals of Life ever more cogent and intimidating. For example, if one says, as I did a moment ago, that the Science failed to respond adequately to the challenge posed by the virus, the answer will be heard right away: we need to do more. That is to say, we need to comply with the commands given by the Science ever more strictly (more lockdowns, more vaccination, and so forth) to comply with the order of Life ever more strictly. And if, by any chance, the Science cannot issue any specific commands in the face of new potential perils looming over us, the order of Life can nonetheless be restated without a pause by denouncing our lack of awareness: we need to do more, even though the Science is not in a position to tell us exactly what is to be done. Even in that case, there is reason to sound the alert.

This alert has been sounded by the Science of Life for decades now: the struggle for life takes no prisoners. When Mother Nature knocks on one's door, the alternative is between survival and extinction. *Tertium non datur.* And it is on the grounds of such a view of Life that a technology of behavior control—to which the theory and practice of preparedness are conducive—can be widely applied. As Skinner put it:

> We have the physical, biological, and behavioural technologies needed "to save ourselves;" the problem is how to get people to use them. It may be that "utopia has only to be willed," but what does that mean? What are the principal specifications of a culture that will survive because it induces its members to work for its survival? . . .
>
> A literature of freedom may inspire a sufficiently fanatical opposition to controlling practices to generate a neurotic if not psychotic response. There are signs of emotional instability in those who have been deeply affected by the literature. . . .
>
> There is a kind of natural morality in both biological and cultural evolution.[56]

Skinner's phrase "a kind of natural morality" brings to light the naturalistic fallacy lying at the heart of the metaphysics of Life and of the current biopolitical technologies of power. In brief, it is thought that "survival is the only value according to which a culture is eventually to be judged,"[57] and that the Science of Life should be charged with the task of watching over the enforcement of the laws of nature.[58]

During the COVID-19 pandemic, the equivalence between being naturally impaired and being morally reprehensible got maximum visibility. In the early months, sick individuals were often suspected of having violated lockdowns or mask mandates, particularly when the epidemiological curve was growing fast. Then, after the start of the inoculation campaign, those who tested positive were often suspected of being antivaxxers. In both cases, transgressors—whether sick or infected—were convicted of a moral crime. And the association between health and morality continued to be made all along. Thus, the Western world began to resemble the setting of Samuel Butler's *Erewhon*, an imaginary distant land where sickness is judged the greatest sin, whereas being in good shape is considered a moral virtue:

> This is what I gathered. That in that country if a man falls into ill health, or catches any disorder, or fails bodily in any way before he is seventy years old, he is tried before a jury of his countrymen, and if convicted is held up to public scorn and sentenced more or less severely as the case may be. There are subdivisions of illnesses into crimes and misdemeanours as with offences amongst ourselves—a man being punished very heavily for serious illness, while failure of eyes or hearing in one over sixty-five, who has had good health hitherto, is dealt with by fine only, or imprisonment in default of payment.[59]

It is well known that Butler disliked Darwin's ideas and loved making fun of them. In his witty description of Erewhonian mores, his distrust of the Darwinian morals of Life comes out very clearly. For him, Erewhon is the dystopian society that may arise one day if our view of nature and human beings remains anchored in Darwin's, and if we subordinate all aspects of human existence to Life and survival. Interestingly, Butler's clairvoyance did not pass unnoticed to Skinner. As he pointed out in his essay on freedom and dignity, "Our culture differs from the Erewhon of Samuel Butler in imposing no punitive sanctions on illness."[60] But the fact remains that in a society like ours, in which the morals of Life set the agenda, illness is to be judged morally wrong, even when it is not sanctioned with punitive measures.

Obviously, even though Butler was well ahead of his time and able to highlight some nontrivial implications of Darwin's theses—he was the first to realize, for example, that Darwinian evolution may lead to Artificial Life and a world inhabited by living machines that will supplant human beings—as well as some arduous problems raised by the theory of natural selection (in particular, that of teleology), he could not foresee events that were very far away. In his day, he could not suspect that the very meaning of "health" was going to change.[61]

For centuries, the concept of health has been used to connote the condition of a human body, the organism of an individual that may or may not be affected by illness; and it is still in this sense that the word is used by Butler and the people of Erewhon. This is not, however, the sense in which we understand health nowadays, in that we look at the human body through the lens of another body, the population's body, the disembodied body to which epidemiology and other branches of the Science of Life give access. When attention is drawn to this body, two things happen: first, the human body dissolves into "a life," a fragment of the population's body; second, health becomes a question of survival and extinction.

Starting from the bottom, by prioritizing a biological-epidemiological approach to the living focused on populations over a medical approach focused on the individual body and its various impairments, often related to each other, it becomes increasingly difficult to conceive of health in terms of well-being, or to think of illness as something that diminishes the individual's well-being and causes death only under highly complex and serious circumstances. Instead, one is likely to identify all pathological conditions with so many threats to survival, no matter what the nature of the disease under scrutiny might be, for the correlation between morbidity and mortality rates continues to be the paramount parameter for the epidemiological evaluation of diseases.[62] On a population scale, illness is ultimately all that poses a threat to survival in one way or another, whereas health is all that increases chances of survival.

This dramatization of illness and health is not without consequences for "a life." The first is that every disease may be perceived as a question of life and death, especially when it is framed as a population disease. This partly explains the overreaction and mass hysteria triggered by the COVID-19 pandemic. As already remarked, throughout the pandemic, no clear-cut distinction was made between being infected and becoming ill, and this caused a growing anxiety among the citizenry. Moreover, the danger was exaggerated because of hasty overgeneralizations that underestimated the difference between the young and the old, or between healthy individuals and those who were truly at risk due to obesity and other physical impairments.

In the end, such developments attest that a new understanding of the health of "a life" has become commonplace over the years, obfuscating the public's perception of what it means to be healthy or sick. A threat to health is now considered, more often than not, a threat to survival. That is, when "a life" is struck by illness, even if the illness in question is not so severe and worrying, "a life" feels like being in mortal danger—and it feels that way particularly when the disease is caused by an epidemic spreading to the whole population.

In this manner, a parallel between "a life" and the population is drawn, and this parallel reveals itself to be more than a mere analogy: by virtue of mythic metonymy, it turns into a distressing identity. The needs of "a life" are identified with the needs of the whole population. As a result, it is believed that "a life" and the population respond to the same treatment, with no exceptions. With mass vaccination against COVID-19, this belief was carried to extremes. Despite the fact that messenger ribonucleic acid (mRNA) vaccines had not been tested for a sufficient lapse of time and nobody knew about side effects in the long term, vaccination was prescribed to everybody. The possibility that the needs of a certain human body differ from the needs of the population's body was ruled out. The assumption was that there cannot be any significant discrepancy, not to say patent contradiction, between the two levels of intervention. As time went by, this assumption was raised to the dignity of dogma. The Science of Life strenuously defended this dogma from all criticism, overlooking or denying the remarkable amount of evidence that had grown in the meantime and spoke against it. For the technocratic and scientific elites as well as the vast majority of citizens, the human body had to merge into the population's body and decompose into "a life," the singular of the plural noun "population," regardless of the unpredictable consequences that this scheme might have for certain individuals.

None of this can be understood, as should be clear by now, unless one takes into account the moral overtones of the identification of "a life" with the population. Such overtones explain the harshness with which any opposition to the biopolitical management of the pandemic was suppressed. As noted earlier, from the moment when emergency measures such as lockdowns and mask mandates were adopted by governments, most people looked askance at those who got infected, for the latter were suspected of having transgressed against the order of Life. Nothing changed after the start of the inoculation campaign. If anything, the moral burden on "a life" became even heavier. Those who did not want, or simply hesitated, to get vaccinated were firmly condemned by the authorities, the media, and the rest of the population. At the beginning, the reason for this public reprimand was that a high vaccination rate was deemed necessary to contain the spread of the pandemic. But shortly after, when it was discovered that COVID-19 vaccines did not prevent anyone from getting infected or infecting others, the stigmatization of so-called antivaxxers did not stop or calm down.

Under such circumstances, the moral burden on individuals could only become disproportionate. Since the beginning of the inoculation campaign, pressure to get vaccinated and boosted with products that were still under clinical evaluation, and whose evaluation was partly shrouded by secrecy, had already undermined some of the ethical principles established by the Nuremberg Code. But when it became evident that vaccination had no significant effect on the infection rate, such pressure started to erode some basic constitutional principles as well, in particular the principle of voluntary treatment, according to which one cannot be forced by any means to receive a medical treatment unless one lacks the capacity to make an informed and voluntary decision. Public scorn, vaccine certificates (Green Passes), and vaccine mandates for some professional categories blatantly contradicted this principle. The unintended consequence of all this was that the reprobates were obliged to follow the orders issued by the Science of Life against their own will and were thus treated de facto as mentally ill or handicapped. Skinner would have been thrilled to see that his ideas had become effective on such a large scale.

To conclude, it is worth returning to the issue of the "autonomous man" one more time. According to Skinner, human freedom is nothing other than a literary myth. The distance between him and Immanuel Kant, the philosopher who did the most to substantiate the notion of moral autonomy, could not be greater. For the latter, nobody can say whether human freedom exists. Nobody can prove that freedom exists, but nobody can prove the opposite either, because human freedom cannot be the subject of scientific investigation and true knowledge. Doubt and confusion reign supreme on this matter. That being said, doubt and confusion are more than enough reason to hold firm to human autonomy. Thus, Kant argues, the door is wide open for a "metaphysics of morals" hinged on the idea of human autonomy. Indeed, freedom and personal responsibility cannot be posited or denied from a theoretical point of view, but freedom and responsibility can nonetheless be assumed from the point of view of practical reason: I am free to think that I am free every time I take action because nobody can demonstrate the contrary. It follows that freedom is, as Kant goes on to explain, "the only one innate right," the only right that can be considered inherent in all rational beings when they act and interact with each other: freedom is their primary right—a metaphysical right, according to Kant—not because rational beings can *know that* they are free, but only because nobody can refute human freedom, and therefore nobody can be denied freedom. All other rights are secondary and derive from that one innate right, which finds its origin in a metaphysical riddle that no human being is in a position to unravel. It is on this right, both indemonstrable and irrefutable, that modern civilization is practically, historically, grounded:

> *Freedom* (independence from being constrained by another's choice), insofar as it can coexist with the freedom of every other in accordance with a universal law, is the only original right belonging to every man by virtue of his humanity. This principle of innate freedom already involves authorizations, which are not really distinct from it (as if they were members of the division of some higher concept of a right): innate *equality*, that is, independence from being bound by others to more than one can in turn bind them; hence a man's quality of being *his own master* (*sui iuris*), as well as being a man *beyond reproach* (*iusti*), since before he performs any act affecting rights he has done no wrong to anyone; and finally, his being authorized to do to others anything that does not in itself diminish what is theirs, so long as they do not want to accept it.[63]

In all likelihood, Hegel had in mind not only himself but also Kant when he famously claimed that philosophy amounts to comprehending one's own time in thoughts.[64] Kant is in fact the philosopher who gave us the most accurate picture of the metaphysical foundations of modern political societies. If the modern age is "the age of rights," as Norberto Bobbio put it, then the modern age is the era of human freedom, the era of the "autonomous man" who is able to claim and obtain rights, to wonder each time whether he has thus achieved freedom, and to keep always open the *possibility* of freedom, which cannot be rejected on a theoretical level but needs to be reaffirmed on a practical, political level over and over again.[65] For this reason, even Hegel, who took a step beyond Kant to argue that the notion of moral autonomy should be contextualized and combined with the notion of *Sittlichkeit* (ethical order), restated the concept: "The Idea of right is freedom."[66] So long as we live in modern societies, freedom remains a possibility, a promise, a project.

Every time this possibility is ruled out, therefore, modern societies are shaken to their foundations, which are metaphysical in nature. And this is exactly what we are witnessing these days, and what Skinner had theorized in his writings. With his portrait of a future society "beyond freedom and dignity," Skinner was making a plea for nothing less than an overcoming of modernity. For him, who regarded himself as a fierce defender of science, to overcome modernity means, first and foremost, to overcome metaphysics—the metaphysics of moral autonomy. But does metaphysics come to an end following Skinner's program? In reality, the result is that another set of metaphysical assumptions deep-seated in modern thought and science—the metaphysics of Life—become hegemonic to the detriment of the "autonomous man," thereby eroding the foundations of modern civilization. Today, as we have been told several times during and after the pandemic, we are on the way to the "new normal." This concept has nothing to do with the old norms and normality, about which Foucault spoke in his works. Rather, the "new normal" is the watchword for a process of remythologization and resacralization of power.

This is not the first time that the metaphysics of Life has taken center stage on a political level and started releasing its effects on the whole of society, pushing the metaphysics of moral autonomy to the margins. Something of the sort had already happened in Nazi Germany less than a century ago, and perhaps this is the reason why some are tempted to overemphasize the similarities between the past and the present.[67] As a matter of fact, however, the two situations remain incomparable from a sociopolitical point of view, save for one characteristic: in both cases, the metaphysics of Life fosters an all-pervasive *ideology* that admits no exception and makes it impossible to conceive of any other way of addressing the problems with which society is confronted. At that point, ideology turns into a sociopolitical *mythology*, as it can no longer be challenged by any competing system of beliefs.

This transition from ideology to mythology was apparent in Nazi Germany.[68] In today's world, the transition is less easy to detect; nevertheless, the scientific ideology that raised its voice during the pandemic, based as it is on the rejection of any alternative, does fuel a kind of mythology. This mythology, that should allow human beings to find their place in the cosmos, is changing our society and culture in depth. It is probably also changing the very meaning of the word "culture." It remains to be seen whether it is conducive to an endurable form of life. In fact, transmodernity seems to put too heavy a burden on "a life."

Thanks to the mythic metonymy that defines transmodern times, "a life" now bears the responsibility for the health, that is the survival, of the entire population to which it belongs, and with which it identifies without noticing. In our age, the difference between the health of anyone and the salvation of the population at large tends to be annulled. The Latin word *salus* conveys both concepts, that represent the two sides of the *pars qua totum* identity around which our world revolves: *to heal* "a life" means *to save* Life as such.

Seen in this light, the transmodern mythology of Life looks like a religion of *salus*, that compensates for the decline of the Christian doctrines of salvation. But significantly, within the new dogmatic framework, the deity, the mythic giver and receiver of *salus*, is no longer a transcendent divinity. Life, unlike the Christian God, is totally immanent, as it has no existence outside living beings—even though no living being can say "I am the life," because "I," the embodied individual, evaporates here, together with the moral person.

In the place of the "I," there emerges "a life," affected by a feeling of death-in-life that can only grow over time. This feeling is partly because "a life" swings between being and nonbeing in its perennial endeavor to reach salvation, and partly because Life itself, by condemning "a life" to such ontological precariousness, condemns itself to the same fate. *Pars qua totum*: Life itself, being immanent in "a life," swings between being and nonbeing. Life itself struggles for existence in each of us. How long this struggle will last depends on how long we will stand the ontological instability that it causes. The day that we lose our balance, we will cease to adhere to the metaphysical assumptions upon which the Science of Life is based.

Notes

Preface

1. Radu Popa, quoted in Carl Zimmer, *Life's Edge: The Search for What It Means to Be Alive* (New York: Dutton, 2021), 272.

2. Carol Cleland, *The Quest for a Universal Theory of Life: Searching for Life as We Don't Know It* (Cambridge: Cambridge University Press, 2010).

3. See Paul-Antoine Miquel, *Qu'est-ce que la vie?* (Paris: Vrin, 2007): "Qu'est-ce que la vie? Dois-je avouer que je trouve la question mauvaise et mal posée? Je voudrais qu'on en finisse avec ce type de questionnement essentialiste. *La vie en effet n'est rien*" (86).

4. This quote is attributed to John von Neumann in the literature, although no specific source is ever mentioned. For more details, see John von Neumann, "The General and Logical Theory of Automata," in *Collected Works, Volume V: Design of Computers, Theory of Automata and Numerical Analysis*, ed. A. H. Taub (Oxford, UK: Pergamon Press, 1963), 288–328.

5. See astrobiology.nasa.gov/research/life-detection/about.

6. John Maynard Smith, *The Problems of Biology* (Oxford: Oxford University Press, 1986), 7.

7. There are some noteworthy exceptions. To name only two: Lynn Margulis, *The Symbiotic Planet: A New Look at Evolution* (London: Phoenix, 1999); Didier Raoult, *Dépasser Darwin* (Paris: Plon, 2010).

8. Davide Tarizzo, *Life: A Modern Invention* (Minneapolis: University of Minnesota Press, 2017).

Introduction

1. Ernest Becker, *The Denial of Death* (New York: Free Press, 1973). For an introduction to terror management theory, see Sheldon Solomon, Jeff Greenberg, and Tom Pyszczynski, *The Worm at the Core: On the Role of Death in Life* (New York: Random House, 2015).

2. Tom Pyszczynski, Sheldon Solomon, and Jeff Greenberg, *In the Wake of 9/11: The Psychology of Terror* (Washington, DC: American Psychological Association, 2003), 13–14.

3. Pyszczynski et al., *In the Wake of 9/11*, 16.

4. Tom Pyszczynski, McKenzie Lockett, Jeff Greenberg, and Sheldon Solomon, "Terror Management Theory and the COVID-19 Pandemic," *Journal of Humanistic Psychology* 61, no. 2 (2021): 173.

5. Pyszczynski et al., "Terror Management Theory and the COVID-19 Pandemic": 178.

6. For more detail, see, for instance, Phyllis Palgy and Henry Abramovitch, "Death: A Cross-Cultural Perspective," *Annual Review of Anthropology* 13 (1984): 385–417.

7. Philippe Ariès, *Western Attitudes Toward Death: From the Middle Ages to the Present* (London: Marion Boyars, 1976), 25.

8. Ariès, *Western Attitudes Toward Death*, 85.

9. Robert Hertz, "A Contribution to the Study of the Collective Representation of Death," in *Death and the Right Hand* (London: Routledge, 2004), 27–28.

10. See Davide Tarizzo, *Life: A Modern Invention* (Minneapolis: University of Minnesota Press, 2017).

11. On Aristotle and Darwin, see Étienne Gilson, *From Aristotle to Darwin and Back Again: A Journey in Final Causality, Species, and Evolution* (Notre Dame, IN: University of Notre Dame Press, 1984).

12. John Dewey, *The Influence of Darwin on Philosophy, and Other Essays in Contemporary Thought* (New York: Henry Holt and Company, 1910), 1.

13. Karl Popper, *Unended Quest: An Intellectual Autobiography* (London: Routledge, 2002), 194–210.

14. "Nobody to date has yet found a demarcation criterion according to which Darwin can be described as scientific." Imre Lakatos and Paul Feyerabend, *For and Against Method* (Chicago: University of Chicago Press, 1999), 24.

15. Jerry Fodor and Massimo Piattelli-Palmarini, *What Darwin Got Wrong* (New York: Farrar, Straus and Giroux, 2010), xiv.

16. On the axiological implications of the modern understanding of Life, see also Hans Jonas, *The Imperative of Responsibility: In Search of an Ethics for the Technological Age* (Chicago: University of Chicago Press, 1984), 79.

17. Michel Foucault, *The History of Sexuality. Volume I: An Introduction* (New York: Pantheon Books, 1978), 140.

18. Foucault, *The History of Sexuality. Volume I*, 149–150.

19. Tarizzo, *Life: A Modern Invention*, 185–220. See also Martin Heidegger, *Nietzsche. Volumes III and IV* (New York: HarperCollins, 1991); György Lukács, *The Destruction of Reason* (Atlantic Highlands, NJ: Humanities Press, 1981); Richard Weikart, *From Darwin to Hitler: Evolutionary Ethics, Eugenics, and Racism in Germany* (New York: Palgrave, 2004); Enzo Traverso, *The Origins of Nazi Violence* (New York: New Press, 2003).

20. Michel Foucault, *The Order of Things: An Archaeology of the Human Sciences* (London: Routledge, 1989), 139.

21. Samuel Coleridge, *Hints Towards the Formation of a More Comprehensive Theory of Life* (London: Churchill, 1848), 21–22.

22. The creative power of natural selection and evolution has often been emphasized by biologists, who regard it as a feature of the biotic world as distinct from the abiotic world. Sometimes, this feature is extended to the whole universe. See, for instance, Stuart A. Kauffman, *Humanity in a Creative Universe* (Oxford: Oxford University Press, 2016).

23. Fodor and Piattelli-Palmarini, *What Darwin Got Wrong*, 121.

24. Fodor and Piattelli-Palmarini, *What Darwin Got Wrong*, 149.

25. As Ernst Mayr put it, "Natural selection does its best to favor the production of [genetic] codes guaranteeing behavior that increases fitness. A behavior program that guarantees instantaneous correct reaction to a potential food source, to a potential enemy, or to a potential mate will certainly give greater fitness in the Darwinian sense than a program that lacks these properties." Ernst Mayr, "Cause and Effect in Biology: Kinds of Causes, Predictability, and Teleology as Viewed by a Practicing Biologist," *Science* 134, no. 3489 (1961): 1504. The phrase "greater fitness in the Darwinian sense" does not mean better adaptation; it means greater reproductive success. This is why, as Mayr adds a few lines later, "the theory of natural selection can describe and explain phenomena with considerable precision, but it cannot make reliable predictions, except through such trivial and meaningless circular statements as, for instance: 'the fitter individuals will on the average leave more offspring'" (1504).

26. In this regard, it is worth recalling the words of one of the best-known neo-Darwinian manifestos: "There is, of course, nothing conscious or intentional in the action of natural selection. . . . Natural selection is at one and the same time a blind and a creative process." Theodosius Dobzhansky, "Nothing in Biology Makes Sense Except in the Light of Evolution," *American Biology Teacher* 35, no. 3 (1973): 127.

27. The difference between the niche and the habitat is crucial for today's ecology. In a nutshell, no two species can occupy the same niche, the definition of which depends on species-specific traits, whereas two or more species can live in the same habitat. It follows that the niche evolves. "If the niche depends on traits and the traits are shaped by evolution, the obvious corollary is that the niche must depend on evolution." Jonathan M. Chase and Mathew A. Leibold, *Ecological Niches: Linking Classical and Contemporary Approaches* (Chicago: University of Chicago Press, 2003), 161.

28. Charles Darwin, *On the Origin of Species by Means of Natural Selection* (London: John Murray, 1859), 66–67.

29. The idea that living beings are "*objects endowed with a purpose or project*," Monod argues, "is essential to the very definition of living beings. We shall maintain that the latter are distinct from all other structures or systems present in the universe through this characteristic property, which we shall call *teleonomy*. . . . It will be readily seen that, in this or that species situated higher or lower on the animal scale, the achievement of the fundamental teleonomic project (i.e., invariant reproduction) calls assorted, more or less elaborate and complex structures and performances into play. The fact must be stressed that concerned here are not only the activities directly bound up with reproduction itself, but all those that contribute—be it very indirectly—to the species' survival and multiplication." Jacques Monod, *Chance and Necessity: An Essay on the Natural Philosophy of Modern Biology* (New York: Vintage Books, 1972), 9, 15.

30. Richard Dawkins, *The Selfish Gene* (Oxford: Oxford University Press, 1976).

31. Niles Eldredge, *Reinventing Darwin: The Great Debate at the High Table of Evolutionary Theory* (New York: John Wiley & Sons, 1995), 40.

32. "We remember the example of the camera: if we agree that this object's existence and structure realize the 'project' of capturing images, we must also agree, obviously enough, that a similar project is accomplished with the emergence of the eye of a vertebrate. But it is only as a part of a more comprehensive project that each individual project, whatever it may be, has any meaning. All the functional adaptations in living beings, like the artifacts they produce, fulfill particular projects which may be seen as so many aspects or fragments of a unique primary project, which is the preservation and multiplication of the species. To be more precise, we shall arbitrarily choose to define the essential teleonomic project as consisting in the transmission from generation to generation of the invariance content characteristic of the species. All the structures, all the performances, all the activities contributing to the success of the essential project will hence be called 'teleonomic'." Monod, *Chance and Necessity*, 14.

33. The notion of a primary project or purpose revives the metaphor of design within the framework of neo-Darwinian selectionism. As Michael Ruse remarks, "Whether or not organisms really are designed, thanks to natural selection, they (or rather, they inasmuch as they are adaptive or adapted) seem as if designed (for the ends of survival and reproduction). We may no longer be thinking of a literal designer up there in the sky, but the mode of understanding persists. For the natural theologian, the heart is literally designed by god—metaphorically, we compare it to a pump made by humans. For the Darwinian, the heart is made through natural selection, but we continue, metaphorically, to understand it as made by humans." Michael Ruse, *Darwin and Design: Does Evolution Have a Purpose?* (Cambridge, MA: Harvard University Press, 2003), 265.

34. Georges Canguilhem, *The Normal and the Pathological* (New York: Zone Books, 1991), 201.

35. Claude Lévi-Strauss, *The Savage Mind* (Chicago: University of Chicago Press, 1966). *La pensée sauvage* is the original French title.

36. Foucault, *The Order of Things*, 303.

Chapter 1

1. Paul Barrett, Peter J. Gautrey, Sandra Herbert, David Kohn, and Sydney Smith, eds., *Charles Darwin's Notebooks, 1836–1844* (Cambridge: Cambridge University Press, 1987), C104.

2. The word "biology" was introduced independently by various thinkers. In the third volume of his *Philosophia Naturalis* of 1766, Michael Christoph Hanov defines the part of physics that studies organic beings as "biologia." In 1799, the English physician Thomas Beddoes uses the word "biology" in his *Contributions to Physical and Medical Knowledge: Principally from the West of England*. One year later, in a footnote to his *Propedeutik zum Studium der gesammten Heilkunde*, the German physiologist Karl Friedrich Burdach speaks of "Biologie." But it is only in 1802, with the works of Gottfried Reinhold Treviranus (*Biologie oder Philosophie der lebenden Natur*) and Jean-Baptiste Lamarck (*Hydrogéologie*), that the term came into its modern scientific usage. For more details on the history of the word "biology," see Arthur Meyer, "Karl Friedrich Burdach and His Place in the History of Neuroanatomy," *Journal of Neurology, Neurosurgery, and Psychiatry* 33, no. 5 (1970): 553–561; William Coleman, *Biology in the Nineteenth Century: Problems of Form, Function, and Transformation* (Cambridge: Cambridge University Press, 1977); Marjorie Grene and David Depew, *The Philosophy of Biology: An Episodic History* (Cambridge: Cambridge University Press, 2004).

3. John Maynard Smith, *The Problems of Biology* (Oxford: Oxford University Press, 1986), 7.

4. Jean Gayon, *Darwin et l'après-Darwin: Une historie de l'hypothèse de la sélection naturelle* (Paris: Kimé, 1992), 190.

5. As Dobzhansky remarks, "The unity of life is no less remarkable than its diversity . . . Seen in the light of evolution, biology is, perhaps, intellectually the most satisfying and inspiring science. Without that light it becomes a pile of sundry facts—some of them interesting or curious but making no meaningful picture as a whole . . . It is remarkable that more than a century ago Darwin was able to discern so much evolution without having available to him the key facts discovered since." Theodosius Dobzhansky, "Nothing in Biology Makes Sense Except in the Light of Evolution," *American Biology Teacher* 35, no. 3 (1973): 127, 129.

6. Improvement is another key concept in Darwin's theory. In his magnum opus, he writes: "It may metaphorically be said that natural selection is daily and hourly scrutinising, throughout the world, every variation, even the slightest; rejecting those that are bad, preserving and adding up all that are good; silently and insensibly working, *whenever and wherever opportunity offers*, at the improvement of each organic being in relation to its organic and inorganic conditions of life. We see nothing of these slow changes in progress, until the hand of time has marked the long lapse of ages, and then so imperfect is our view into long past geological ages, that we only see that the forms of life are now different from what they formerly were." Charles Darwin, *The Origin of Species*, 6th ed. (Cambridge: Cambridge University Press, 2009 [orig. 1876]), 65–66.

7. Darwin, *The Origin of Species*, 429.

8. On the analogy between evolutionary forces and Newtonian forces, see Elliott Sober, *The Nature of Selection: Evolutionary Theory in Philosophical Focus* (Cambridge, MA: MIT Press, 1984); Mohan Matthen and André Ariew, "Two Ways of Thinking about Fitness and Natural Selection," *Journal of Philosophy* 99, no. 2 (2002): 55–83; Denis M. Walsh, "The Pomp of Superfluous Causes: The Interpretation of Evolutionary Theory," *Philosophy of Science* 74, no. 3 (2007): 281–303; Tim Lewens, "The Natures of Selection," *British Journal for the Philosophy of Science* 61, no. 2 (2009): 1–21.

9. It is well known that Darwin never ceased to lament that selection "depends on what we in our ignorance call spontaneous or accidental variability." Charles Darwin, *The Variation of Animals and Plants under Domestication*, vol. II (London: John Murray, 1868), 248.

10. The following passage from *The Variation of Animals and Plants under Domestication* encapsulates the meaning of unconscious selection: "The different strains, just alluded to, which have been actually produced by breeders without any wish on their part to obtain such a result, afford excellent evidence of the power of unconscious selection. This form of selection has probably led to far more important results than methodical selection, and is likewise more important under a theoretical point of view from closely resembling natural selection. For during this process the best or most valued individuals are not separated and prevented from crossing with others of the same breed, but are simply preferred and preserved; yet this inevitably leads to their gradual modification and improvement; so that finally they prevail, to the exclusion of the old parent-form." Darwin, *The Variation of Animals and Plants under Domestication*, vol. II, 424.

11. On Darwin's definition of his own theory as "one long argument," see Ernst Mayr, *One Long Argument: Charles Darwin and the Genesis of Modern Evolutionary Thought* (Cambridge MA: Harvard University Press, 1993).

12. Michel Foucault, *The Order of Things: An Archaeology of the Human Sciences* (London: Routledge, 1989), 293.

13. Foucault, *The Order of Things*, 297.

14. Giovanni Pico della Mirandola, *Oration on the Dignity of Man* (Chicago: Henry Regenry, 1956 [orig. 1486]), 6.

15. Johann Gottfried Herder, *Outlines of a Philosophy of the History of Man* (New York: Bergman, 1966 [orig. 1784–1792]), 109, 112, 114.

16. Darwin, *The Origin of Species*, 166.

17. Barrett et al., *Charles Darwin's Notebooks*, E85.

18. Georges-Louis Leclerc de Buffon, *Buffon's Natural History: Containing a Theory of the Earth, a General History of Man, of the Brute Creation, and of Vegetables, Minerals, &c. &c., Vol. X* (London: H. D. Symonds, 1797), 352–353.

19. "The germs of existence contained in this spot of earth, with ample food, and ample room to expand in, would fill millions of worlds in the course of a few thousand years. Necessity, that imperious all pervading law of nature, restrains them within the prescribed bounds. The race of plants and the race of animals shrink under this great restrictive law." Thomas Robert Malthus, *An Essay on the Principle of Population*, 6th ed. (London: Murray, 1826), 5.

20. See, for instance, Ernst Mayr, *Evolution and the Diversity of Life* (Cambridge, MA: Harvard University Press, 1976); Elliott Sober, "Evolution, Population Thinking, and Essentialism," *Philosophy of Science* 47, no. 3 (1980): 350–383.

21. "The initial elementary events which open the way to evolution in the intensely conservative systems called living beings are microscopic, fortuitous, and utterly without relation to whatever may be their effects upon teleonomic functioning. But once incorporated in the DNA structure, the accident—essentially unpredictable because always singular—will be mechanically and faithfully replicated and translated: that is to say, both multiplied and transposed into millions or billions of copies. Drawn out of the realm of pure chance, the accident enters into that of necessity, of the most implacable certainties. For natural selection operates at the macroscopic level, the level of organisms. Even today a good many distinguished minds seem unable to accept or even to understand that from a source of noise natural selection alone and unaided could have drawn all the music of the biosphere. In effect natural selection operates *upon* the products of chance and can feed nowhere else." Jacques Monod, *Chance and Necessity: An Essay on the Natural Philosophy of Modern Biology*, New York: Vintage Books, 1972), 118.

22. Darwin, *The Origin of Species*, 106.

Chapter 2

1. Karl Pearson, *The Grammar of Science*, 3rd ed. (London: Adam and Charles Black, 1911), 415–416.

2. For more details on this definition, see Jean Gayon, *Darwin et l'après-Darwin: Une historie de l'hypothése de la sèlection naturelle* (Paris: Kimé, 1992).

3. Charles Darwin, *The Variation of Animals and Plants under Domestication*, vol. I (London: John Murray, 1868), 8–9.

4. As Michael Ruse points out, "In the physical sciences we think a lot about the future. I am sure that there are today many earth scientists, hunched over their computers, building models about the effects of impacts. But it is prior causes alone—what are often known as 'efficient causes'—to which appeal is made. As one might say, it is turtles all of the way down. Which physical events brought on which physical events and in what fashion, and what are the likelihoods of it all happening again? Likewise, in the case of normal causal processes like physiological development or computer form filling, it is all a matter of efficient causes." Michael Ruse, *On Purpose* (Princeton, NJ: Princeton University Press, 2018), xii–xiii. Regarding the concept of purpose or final cause, Carl Gustav Hempel emphasizes: "When a certain form of mimicry is said to have the purpose of preserving a given species by protecting its members from their natural enemies, no direct or indirect interpretation of 'purpose' in terms of observables is provided; i.e., no criteria are laid down by means of which it is possible to test assertions about the purposes of biological phenomena and to decide, let us say, between the view just mentioned and the alternative opinion that mimicry has the purpose of bringing aesthetic variety into the animate world . . . This use of teleological terminology may be characterized as an illegitimate transfer from contexts of similar grammatical form, for which, however, an empirical interpretation is available, such as 'the safety valves of steam engines have the purpose of preventing explosions.' This sentence can be interpreted by reference to the intentions and beliefs of the designer, which can be ascertained, at least under favorable conditions, by various observational methods." Carl Gustav Hempel, *Fundamentals of Concept Formation in Empirical Science* (Chicago: University of Chicago Press, 1952), 40.

5. To name but a few examples: Ernest Nagel, *The Structure of Science: Problems in the Logic of Scientific Explanation* (New York: Harcourt, Brace & World, 1961); Carl Gustav Hempel, *Aspects of Scientific Explanation and Other Essays in the Philosophy of Science* (New York: Free Press, 1965); Georg Henrik von Wright, *Explanation and Understanding* (London: Routledge & Kegan Paul, 1971); Marjorie Grene, *The Knower and the Known* (Berkeley: University of California Press, 1974); Robert Cummins, "Functional Analysis," *Journal of Philosophy* 72, no. 20 (1975): 741–765.

6. André Ariew, "Teleology," in *Cambridge Companion to the Philosophy of Biology*, ed. David L. Hull and Michael Ruse (Cambridge: Cambridge University Press, 2007), 179. See also David L. Hull, "Philosophy and Biology," in *Contemporary Philosophy: A New Survey. Volume 2: Philosophy of Science*, ed. Guttorm Fløistad (The Hague: Martinus Nijhoff, 1982), 281–316.

7. "Darwin maintained that evolution has no direction." Stephen Jay Gould, *Ever Since Darwin: Reflections in Natural History* (New York: W. W. Norton, 1992), 13.

8. Francisco J. Ayala, "Teleological Explanations in Evolutionary Biology," *Philosophy of Science* 37, no. 1 (1970): 10.

9. Ayala, "Teleological Explanations in Evolutionary Biology," 13.

10. Ernst Mayr, "Teleological and Teleonomic, a New Analysis," in *Methodological and Historical Essays in the Natural and Social Sciences*, ed. Robert S. Cohen and Marx W. Wartofsky (Dordrecht: Springer Netherlands, 1974), 113. See also Colin Pittendrigh, "Adaptation, Natural Selection and Behavior," in *Behavior and Evolution*, ed. Anne Roe and George Gaylord Simpson (New Haven, CT: Yale University Press, 1958), 390–419; George Gaylord Simpson, "Behavior and Evolution," in *Behavior and Evolution*, ed. Anne Roe and George Gaylord Simpson (New Haven, CT: Yale University Press, 1958), 507–535; Jacques Monod, *Chance and Necessity: An Essay on the Natural Philosophy of Modern Biology* (New York: Vintage Books, 1972).

11. Mayr, "Teleological and Teleonomic, a New Analysis," 102.

12. "I agree with Krieger that Mayr's definition of 'teleonomy' is circular, and have said so in the past. His definition is: 'A teleonomic process or behavior is one that owes its goal-directedness to the operation of a program'; and adds, 'A program might be defined as coded or prearranged information that controls a process (or behavior) leading it toward a goal.' So, a process is teleonomic if governed by a program; and if we want to know what kind of program govern teleonomic processes, Mayr says that they are those programs ('coded or prearranged information') that lead the process 'toward a goal.' There is no much enlightment here." Francisco J. Ayala, "Teleological Explanations Versus Teleology," *History and Philosophy of the Life Sciences* 20, no. 1 (1998): 50.

13. Gould, *Ever Since Darwin*, 13.

14. Charles Darwin, *The Autobiography of Charles Darwin, 1809–1882*, ed. Nora Barlow (London: Collins, 1958), 67–68.

15. See, for instance, Michael Ruse, "Darwin's Debt to Philosophy: An Examination of the Influence of the Philosophical Ideas of John F. W. Herschel and William Whewell on the Development of Charles Darwin's Theory of Evolution," *Studies in History and Philosophy of Science* 6, no. 2 (1975): 159–181; M. J. S. Hodge, "The Structure and Strategy of Darwin's 'Long Argument'," *British Journal for the History of Science* 10, no. 3 (1977): 237–246; M. J. S. Hodge, "Darwin's Argument in the Origin," *Philosophy of Science* 59, no. 3 (1992): 461–464.

16. Charles Darwin, "Letter to George Bentham," May 22, 1863, Darwin Correspondence Project (http://www.darwinproject.ac.uk).

17. "Experience having shown us the manner in which one phenomenon depends on another in a great variety of cases, we find ourselves provided, as science extends, with a continually increasing stock of such antecedent phenomena, or causes (meaning at present merely proximate causes), competent, under different modifications, to the production of a great multitude of effects, besides those which originally led to a knowledge of them. To such causes Newton has applied the term *verae causae*." John Frederick William Herschel, *A Preliminary Discourse on the Study of Natural Philosophy* (Cambridge: Cambridge University Press, 2009 [orig. 1830]), 144.

18. Charles H. Pence, "Sir John F.W. Herschel and Charles Darwin: Nineteenth-Century Science and Its Methodology," *Journal of the International Society for the History of Philosophy of Science* 8 (2018): 133.

19. Herschel, *A Preliminary Discourse*, 149–150.

20. Monod, *Chance and Necessity*, 9, 15.

21. Mayr, "Teleological and Teleonomic, a New Analysis," 114.

22. Richard Dawkins, *The Selfish Gene* (Oxford: Oxford University Press, 1976).

23. Niles Eldredge, *Reinventing Darwin: The Great Debate at the High Table of Evolutionary Theory* (New York: Wiley and Sons, 1995), 203.

24. Galileo Galilei, *Discoveries and Opinions of Galileo*, ed. Stillman Drake (New York: Doubleday & Co., 1957), 276–277.

25. Gottfried Wilhelm von Leibniz, *The Monadology*, in Lloyd Strickland, *Leibniz's Monadology: A New Translation and Guide* (Edinburgh: Edinburgh University Press, 2014), 17.

26. The idea that evolution is a mechanistic process has been argued for by evolutionary biologists time and again. As Mayr put it, "The ultimate causes for the efficiency and seeming purposefulness of these living systems were explained by Darwin in 1859. The adaptiveness of these systems is the result of millions of generations of natural selection. This is the mechanistic explanation of adaptiveness." Mayr, "Teleological and Teleonomic, a New Analysis," 108.

27. For more detail, see Stephen Jay Gould, *The Structure of Evolutionary Theory* (Cambridge, MA: Harvard University Press, 2002).

28. Eldredge, *Reinventing Darwin*, 86.

29. Peter Abrams, "Adaptationism, Optimality Models, and Tests of Adaptive Scenarios," in *Adaptationism and Optimality*, ed. Steven Hecht Orzack and Elliott Sober (Cambridge: Cambridge University Press, 2001), 289.

30. "The fact that possession of a trait complex correlates with increased adaptedness or inclusive fitness of its bearers is not sufficient to establish that it is an (evolutionary) adaptation, for that fact leaves unresolved questions regarding the historical pathway by which that trait arose." Richard M. Burian, "Adaptation: Historical Perspectives," in *Keywords in Evolutionary Biology*, ed. Evelyn Fox Keller and Elizabeth A. Lloyd (Cambridge, MA: Harvard University Press, 1992), 11.

31. To Stephen Jay Gould, who asked, "What good is 5 percent of an eye?" and found this question "excellent," Richard Dawkins famously replied: "An ancient animal with 5 per cent of an eye might indeed have used it for something other than sight, but it seems to me at least as likely that it used it for 5 per cent vision. And actually I don't think it is an excellent question. Vision that is 5 per cent as good as yours or mine is very much worth having in comparison with no vision at all. So is 1 per cent vision better than total blindness. And 6 per cent is better than 5, 7 per cent better than 6, and so on up the gradual, continuous series." Richard Dawkins, *The Blind Watchmaker* (New York: W. W. Norton, 1996), 81.

32. For further details on the rhizomatic conception of evolution at the microbiological level, see Didier Raoult and Eugene V. Koonin, "Microbial Genomics Challenge Darwin," *Frontiers in Cellular and Infection Microbiology* 2 (2012): 1–2.

33. Charles Darwin, *The Origin of Species*, 6th ed. (Cambridge: Cambridge University Press, 2009 [orig. 1876]), 404.

34. Dawkins, *The Blind Watchmaker*, 49.

35. "The entire exercise is, however, an achievement in self-deception. A *target* phrase? Iterations that *most resemble* the target? . . . The mechanism of deliberate design, purged by Darwinian theory on the level of organism, has reappeared in the description of natural selection itself, a vivid example of what Freud meant by the return of the repressed." David Berlinski, "The Deniable Darwin," *Commentary* (June 1996), 27–28.

36. Dawkins, *The Blind Watchmaker*, 71–72.

37. Some ecologists "tend to think that a niche can exist even without the organisms that have adapted to it to some extent. It is these ecologists who propagate the concept of vacant (= unoccupied / empty / free / unfilled) niche. In their opinion, the niche is primarily a property of the environment and not of an organism. Therefore, to emphasize the difference from the currently more widespread concept of Hutchinsonian or ecological niche, such a concept is sometimes referred to as either a non-Hutchinsonian or environmental niche. These authors define vacant niche as the resources that are unused by anyone but are potentially usable." Edmundas Lekevicius, "Vacant Niches in Nature, Ecology, and Evolutionary Theory: a Mini-Review," *Ekologija* 55, no. 3–4 (2009): 165.

38. George Evelyn Hutchinson, *The Ecological Theater and the Evolutionary Play* (Princeton, NJ: Princeton University Press, 1964), 36. The difference between adaptability and adaptation reflects Hutchinson's distinction between the "fundamental niche" (comprising all dimensions of the niche hypervolume) and the "realized niche" (the portion of the fundamental niche that is occupied after interactions with other species). For Hutchinson, evolutionary adaptation and the realized niche result from the struggle for existence among competing species: "If no competitors are present, the hypervolume will constitute the *fundamental niche* of the species" (32). See also George Evelyn Hutchinson, "Concluding Remarks," *Cold Spring Harbor Symposium on Quantitative Biology* 22 (1957): 415–427.

39. Hutchinson, *The Ecological Theater*, 27.

40. See, for instance, John Maynard Smith, *Evolution and the Theory of Games* (Cambridge: Cambridge University Press, 1982); Steven Hecht Orzack and Elliott Sober, "Optimality Models and the Test of Adaptationism," *American Naturalist* 143 (1994): 361–380.

41. See, for instance, Carl Gans, "Adaptation and the Form-Function Relation," *American Zoologist* 28 (1988): 681–697.

42. For more detail on this topic and a list of bibliographical references, see Paul Sheldon Davies, *Norms of Nature: Naturalism and the Nature of Functions* (Cambridge, MA: MIT Press, 2001).

43. Egbert Giles Leigh, Jr., "Adaptation, Adaptationism, and Optimality," in *Adaptationism and Optimality*, ed. Steven Hecht Orzack and Elliott Sober (Cambridge: Cambridge University Press, 2001), 363.

44. Johann Friedrich Blumenbach, *Über den Bildungstrieb und das Zeugungsgeschäfte* (Stuttgart: Gustav Fischer, 1971 [orig. 1781]), 62–63.

45. Georges Cuvier, *The Animal Kingdom: Arranged in Conformity with Its Organization*, vol. 1, trans. Edward Griffith (Cambridge: Cambridge University Press, 2012 [orig. 1816]), 5.

46. The "Historical Sketch" has appeared "in one version or another as a preface to every authorized edition of the *Origin* ever published after the second English edition in 1860. The purpose of the sketch was to give a brief history of opinion about the species question as a prelude to Darwin's own independent contribution to the subject. But its provenance is somewhat obscure. Some things are known about its production, such as when it first appeared and what changes were made to it between its first appearance in 1860 and its final form, for the fourth English edition published in 1866. But how it evolved in Darwin's mind, why he wrote it at all, and what he thought he was accomplishing by prefacing it to the *Origin* remain questions that have not been carefully addressed in the scholarly literature on Darwin." Curtis N. Johnson, "The Preface to Darwin's *Origin of Species*: The Curious History of the "Historical Sketch'," *Journal of the History of Biology* 40 (2007): 530–531.

47. Darwin, *The Origin of Species*, xiv.

48. Darwin, *The Origin of Species*, xvi.

49. Darwin, *The Origin of Species*, xvii.

50. Darwin, *The Origin of Species*, xvii.

51. Timothy Lenoir, *The Strategy of Life: Teleology and Mechanics in Nineteenth-Century German Biology* (Chicago: University of Chicago Press, 1989), 263, 267, 270.

52. Darwin, *The Origin of Species*, xvii.

53. George C. Williams, *Adaptation and Natural Selection: A Critique of Some Current Evolutionary Thought* (Princeton, NJ: Princeton University Press, 1966), vii.

54. For more detail, see Burian, "Adaptation: Historical Perspectives," 7–12.

55. Sewall Wright, "The Roles of Mutation, Inbreeding, Crossbreeding and Selection in Evolution," *Proceedings of the Sixth International Congress of Genetics* 1 (1932): 356–366.

56. Ronald Fisher, *The Genetical Theory of Natural Selection* (Oxford, UK: Clarendon Press, 1930), 38.

57. Walter Benjamin, "Theses on the Philosophy of History," in *Illuminations*, ed. Hannah Arendt (New York: Schocken Books, 1968), 257–258.

58. The English translation of these quotes reads: "I am the Spirit of Eternal Negation" and "A part of that force which, always willing evil, always produces good." Johann Wolfgang von Goethe, *Faust I & II*, trans. Stuart Atkins (Princeton, NJ: Princeton University Press, 2014), 36.

59. Egbert Giles Leigh, Jr., *Adaptation and Diversity: Natural History and the Mathematics of Evolution* (San Francisco: Freeman, Cooper & Company, 1971), 270.

60. Marc Kirschner and John Gerhart, "Evolvability," *Proceedings of the National Academy of Sciences* 95, no. 15 (1998): 8420.

61. Dov Ospovat, "Darwin after Malthus," *Journal of the History of Biology* 12, no. 2 (1979): 211.

62. Dov Ospovat, "God and Natural Selection: The Darwinian Idea of Design," *Journal of the History of Biology* 13 (1980): 170–171.

63. Charles Darwin, *The Foundations of the Origin of Species: Two Essays Written in 1842 and 1844*, ed. Francis Darwin (Cambridge: Cambridge University Press, 1909), 95.

64. Charles Darwin, *On the Origin of Species by Means of Natural Selection* (London: John Murray, 1859), 202.

65. Darwin, *On the Origin of Species by Means of Natural Selection*, 206.

66. Jerry Fodor and Massimo Piattelli-Palmarini, *What Darwin Got Wrong* (New York: Farrar, Straus and Giroux, 2010), 152.

67. Edward O. Wilson, *The Meaning of Human Existence* (New York: W. W. Norton, 2014); Michael Ruse, *A Meaning to Life* (Oxford: Oxford University Press, 2019).

68. Richard Dawkins, *The God Delusion* (London: Bantam Press, 2006); Thomas Henry Huxley, *Evidence as to Man's Place in Nature* (London: Williams & Norgate, 1863).

69. For more detail on the creationist-evolutionist debate, see, for instance, Edward J. Larson, *Trial and Error: The American Controversy over Creation and Evolution* (Oxford: Oxford University Press, 2003).

70. For more detail on sociobiology and evolutionary psychology, see, for instance, Philip Kitcher, *Vaulting Ambitions: Sociobiology and the Quest for Human Nature* (Cambridge, MA: MIT Press, 1985), 213–239; John Dupré, *Darwin's Legacy: What Evolution Means Today* (Oxford: Oxford University Press, 2003), 77–98.

71. "A population is not to be identified with the set of organisms in it." Elliott Sober, "Sets, Species, and Evolution: Comments on Philip Kitcher's 'Species'," *Philosophy of Science* 51, no. 2 (1984): 337.

72. For further detail on this point, see Elliott Sober, "Evolution, Population Thinking, and Essentialism," *Philosophy of Science* 47, no. 3 (1980): 350–383; John Dupré, "Natural Kinds and Biological Taxa," *Philosophical Review* 90, no. 1 (1981): 66–90.

73. Michael T. Ghiselin, *Metaphysics and the Origin of Species* (New York: State University of New York Press, 1997). See also David L. Hull, "A Matter of Individuality," *Philosophy of Science* 45 (1978): 335–360; Marc Ereshesfky, "Darwin's Solution to the Species Problem," *Synthese* 175 (2010): 405–425.

74. See, for instance, Richard Dawkins, *The Extended Phenotype: The Gene as the Unit of Selection* (Oxford: Oxford University Press, 1982). It is worth emphasizing that the ontological disintegration of individual organisms, which is one of the keystones of neo-Darwinism, raises difficult questions. One of them reads as follows: "When a population evolves by natural selection, what, if anything, is the entity that does the adapting?" Elliott Sober, *The Nature of Selection: Evolutionary Theory in Focus* (Cambridge, MA: MIT Press), 204. On this issue, see also Ernst Mayr, "The Objects of Selection," *Proceedings of the National Academy of Sciences of the USA* 94 (1997): 2091–2094.

75. With regard to totemism, it has been observed that "it is this direct perception of the *class*, through the individuals, which characterizes the relation between man and the animal or plant, and it is this also which helps us to understand 'this singular thing that is totemism'." Claude Lévi-Strauss, *Totemism* (London: Merlin Press, 1964), 93.

76. Michael Ruse, "The Darwinian Revolution: Rethinking Its Meaning and Significance," *Proceedings of the National Academy of Sciences* 106, Suppl. 1 (2009): 10044.

77. "It can be claimed that wherever religion exists, there exists at least one area in which a certain indistinctness between nature and society is kept in force. Indeed, the element common to all religions is the idea that natural processes depend to some extent on moral order. Disguised under the idea of man's dependence on forces that transcend him—and that are generally located in nature (whether in the skies or in the forest matters little)—religion in reality expresses the idea that nature is dependent on a human order. This is an idea of nature that has it behaving according to rules or morality, a nature that punishes evil with lightning and rewards the righteous with abundance. From experience of the efficacy of the word to which society delegates its authority over itself, society derives the idea that this authority extends to a nature populated by human images." Valerio Valeri, *Classic Concepts in Anthropology* (Chicago: Hau Books, 2018), 213.

78. Silvano Arieti, *Interpretation of Schizophrenia* (New York: Basic Books, 1974), 437.

Chapter 3

1. For a short overview of philosophical debates on naturalism, see Mario De Caro and David Macarthur, "Introduction: The Nature of Naturalism," in *Naturalism in Question*, ed. Mario De Caro and David Macarthur (Cambridge, MA: Harvard University Press, 2004), 1–17.

2. It is not easy to give an exact definition of scientism. Here, I take my cue from Hilary Putnam, *Renewing Philosophy* (Cambridge, MA: Harvard University Press, 1992), ix–x.

3. The demarcation problem has often been discussed by contemporary philosophers, from Karl Popper to Larry Laudan and beyond. For an updated survey, see Massimo Pigliucci and Maarten Boudry, eds., *Philosophy of Pseudoscience: Reconsidering the Demarcation Problem* (Chicago: University of Chicago Press, 2013); Maarten Boudry and Massimo Pigliucci, eds., *Science Unlimited? The Challenges of Scientism* (Chicago: University of Chicago Press, 2017).

4. The distinction between the "scientific image" and the "manifest image" was first made by Wilfrid Sellars in *Science, Perception, and Reality* (New York: Humanities Press, 1963), 1–40. Sellars, however, emphasized the distance between the two images, not the impact that the former can have on the latter. For more details and some critical remarks on Sellars's view, see Bas van Fraassen, "The Manifest Image and the Scientific Image," in *Einstein Meets Magritte: An Interdisciplinary Reflection. The White Book*, ed. Diederik Aerts, Jan Broekaert, and Ernst Mathijs (Dordrecht, Netherlands: Kluwer, 1999), 29–52.

5. See David Baltimore, "Viral RNA-Dependent Polymerase," *Nature* 226 (1970): 1209–1211; Howard M. Temin and Satoshi Mizutani, "RNA-Dependent DNA Polymerase in Virions of Rous Sarcoma

Virus," *Nature* 226 (1970): 1211–1213; Eugene V. Koonin, "Does the Central Dogma Still Stand?" *Biology Direct* 7, no. 1 (2012): 1–7.

6. Michel Morange, "The Central Dogma of Molecular Biology," *Resonance* 14, no. 3 (2009): 236–247.

7. See, for instance, Richard C. Lewontin, Steven Rose, and Leon J. Kamin, *Not in Our Genes: Biology, Ideology, and Human Nature* (New York: Pantheon Books, 1984); Richard C. Lewontin, *Biology as Ideology: The Doctrine of DNA* (Toronto: Anansi Press, 1991).

8. "By the 1950s, the standard view of carcinogenesis became dominated by the somatic mutation theory (SMT), which is rooted in a reductionist and genetic deterministic stance, including the erroneous assumption of a linear causal link between genotype and phenotype. This stance predicates that cells are like tiny computers run by a program. The lack of a robust theoretical background to frame experiments and explanations has hindered the adoption of productive alternatives in the fields of experimental biology and cancer research for decades. . . . For the sake of the current and future credibility in the fields of biology at large and cancer in particular, a rigorous, critical assessment of the odd situation biologists and cancer researchers are facing is in order." Carlos Sonnenschein and Ana M. Soto, "Over a Century of Cancer Research: Inconvenient Truths and Promising Leads," *PLoS Biology* 18, no. 4 (2020): doi.org/10.1371/journal.pbio.3000670. See also Carlos Sonnenschein and Ana M. Soto, *The Society of Cells: Cancer and Control of Cell Proliferation* (New York: Taylor and Francis, 1999); Harold Varmus, "The New Era in Cancer Research," *Science* 312 (2006): 1162–1165.

9. "Never in the history of the gene has the term had as much force in the popular imagination as in recent years, and, accordingly, never has gene talk had more persuasive—that is, rhetorical—power. The invocation of genes has proven demonstrably effective not only in securing funding and promoting research agendas but also (and perhaps even especially) in marketing the products of a rapidly expanding biotech industry. Indeed, the new partnerships between science and commerce that are daily being forged by the promises of genomics bind genetics to the market with a strength and intimacy that is unprecedented in the annals of basic research in the life sciences. The closer such ties, the greater the research scientist's investment becomes in the rhetorical power of a language that works so well. Such connections are obvious. But less obvious are the ways in which efficacy as tools of persuasion reinforces the perceived value of gene talk in more immediate experimental contexts, and vice versa. Resonances between and among these different effects make it exceedingly difficult to give up on gene talk, in the laboratory as in the marketplace. One might say that such forms of mutual reinforcement are just what make the terminology of scientific practice self-stabilizing, at least in the short run." Evelyn Fox Keller, *The Century of the Gene* (Cambridge, MA: Harvard University Press, 2000), 143–144. See also Barry Barnes and John Dupré, *Genomes and What to Make of Them* (Chicago: University of Chicago Press, 2008).

10. A recent example of this approach can be found in Robert Plomin, *Blueprint: How DNA Makes Us Who We Are* (Cambridge, MA: MIT Press, 2018). For a critical review of this essay, see Nathaniel Comfort, "Genetic Determinism Rides Again," *Nature* 561 (2018): 461–463.

11. See Richard J. Herrnstein and Charles Murray, *The Bell Curve: Intelligence and Class Structure in American Life* (New York: Simon & Schuster, 1994); Russell Jacoby and Naomi Glauberman, eds., *The Bell Curve Debate: History, Documents, Opinions* (New York: Random House, 1995); Stephen Jay Gould, *The Mismeasure of Man* (New York: W. W. Norton, 1996).

12. Karl Kerényi, *Dionysos: Archetypal Image of Indestructible Life* (Princeton, NJ: Princeton University Press, 1976), xxxiv.

13. See Giorgio Agamben, *Homo Sacer: Sovereign Power and Bare Life* (Stanford, CA: Stanford University Press, 1998).

14. Kerényi, *Dionysos*, xxxvii.

15. Kerényi, *Dionysos*, xxxiv.

16. Kerényi, *Dionysos*, 388.

17. See Davide Tarizzo, *Homo Insipiens. La filosofia e la sfida dell'idiozia*, Milan: Franco Angeli, 2004.

18. See Immanuel Kant, *Critique of the Power of Judgment*, ed. Paul Guyer (Cambridge: Cambridge University Press, 2000 [orig. 1790]), 233–234.

19. Karen Neander, "The Teleological Notion of 'Function'," *Australasian Journal of Philosophy* 69, no. 4 (1991): 456–457.

20. There are plenty of examples of such attempts in the history of philosophy and biology. See, for instance, Georg Henrik von Wright's efforts to distinguish teleological explanations from "quasi-teleological" ones that "play a great role in the biological sciences." Georg Henrik von Wright, *Explanation and Understanding* (London: Routledge & Kegan Paul, 1971), 59.

21. Neander, "The Teleological Notion of 'Function'," 463.

22. As soon as the metaphor is neutralized, in fact, natural selection turns into "a passive reflection of 'what worked better than what'." Niles Eldredge, *Reinventing Darwin: The Great Debate at the High Table of Evolutionary Theory* (New York: Wiley and Sons, 1995), 203.

23. See Cruz González-Ayesta, "Scotus' Interpretation of the Difference between *Voluntas ut Natura* and *Voluntas ut Voluntas*," *Franciscan Studies* 66 (2008): 371–412.

24. Ernst Mayr has famously argued that "evolution as such" is one of the five theories (or conjectures) that compose Darwin's theory of natural selection. But he has also stressed that "evolution as such is no longer a theory for a modern author. It is as much a fact as that the earth revolves around the sun rather than the reverse. The changes documented by the fossil record in precisely dated geological strata are the fact that we designate as evolution. It is the factual basis on which the other four evolutionary theories rest. For instance, all the phenomena explained by common descent would make no sense if evolution were not a fact." Ernst Mayr, *What Makes Biology Unique? Considerations on the Autonomy of a Scientific Discipline* (Cambridge: Cambridge University Press, 2004), 100.

25. Edoardo Boncinelli, "La genetica dell'evoluzione," *Le Scienze* 486 (2009): 46. As Mayr put it succinctly, "adaptive means simply: being the result of natural selection." Ernst Mayr, "Teleological and Teleonomic, a New Analysis," in *Methodological and Historical Essays in the Natural and Social Sciences*, ed. Robert S. Cohen and Marx W. Wartofsky (Dordrecht: Springer Netherlands, 1974), 107.

26. The "conditions of life," usually understood as "final causes" by nineteenth-century naturalists, were an issue of primary importance for Darwin: "Naturalists continually refer to external conditions, such as climate, food, &c., as the only possible cause of variation. In one very limited sense, as we shall hereafter see, this may be true; but it is preposterous to attribute to mere external conditions, the structure, for instance, of the woodpecker, with its feet, tail, beak, and tongue, so admirably adapted to catch insects under the bark of trees. . . . Seedlings from the same fruit, and the young of the same litter, sometimes differ considerably from each other, though both the young and the parents, as Müller has remarked, have apparently been exposed to exactly the same conditions of life; and this shows how unimportant the direct effects of the conditions of life are in comparison with the laws of reproduction, and of growth, and of inheritance; for had the action of the conditions been direct, if any of the young had varied, all would probably have varied in the same manner." Charles Darwin, *On the Origin of Species by Means of Natural Selection* (London: John Murray, 1859), 3, 10.

27. Boncinelli, "La genetica dell'evoluzione," 50.

28. "It is important to be clear about what is being asserted by the central dogma. It is not true that DNA can replicate without proteins: enzymes are needed. Further, changes in enzymes can alter the way in which a particular DNA sequence is translated. What does seem to be true, however, is that, if a protein with a new amino-acid sequence is present in a cell, it cannot cause the production of a DNA molecule with the corresponding base sequence. Notice that this is not a logical necessity. Machines that translate information can sometimes work both ways: a tape recorder can translate sounds into magnetic patterns on a tape, and vice versa. But some machines translate only in one direction: you cannot cut a record by singing into the speaker of a record-player. The central dogma claims that the relation between nucleic acids and proteins resembles a record-player, and not a tape recorder. . . . If the central dogma is true, and if it is also true that nucleic acids are the only means whereby information is transmitted between generations, this has crucial implications for evolution." John Maynard Smith, *Evolutionary Genetics* (Oxford: Oxford University Press, 1998), 9–10.

29. "It is remarkable that one can formulate principles such as the Sequence Hypothesis and the Central Dogma, which explain many striking facts and yet for which proof is completely lacking. This gap between theory and experiment is a great stimulus to imagination." Francis H. Crick, "On Protein Synthesis," *Symposia of the Society for Experimental Biology* 12 (1958): 160–161.

30. According to Dawkins, the central dogma is "one of the great achievements of modern biology." Richard Dawkins, *The Extended Phenotype: The Gene as the Unit of Selection* (Oxford: Oxford University Press, 1982), 14.

31. See Immanuel Kant, *Critique of Pure Reason*, ed. Paul Guyer and Allen W. Wood (Cambridge: Cambridge University Press, 1998 [orig. 1781–1787]), 384–387.

32. This idea has been advocated by a number of philosophers in recent years. One of them begins his essay on the topic with the following statements: "This book has two interwoven theses. The first concerns science. It is the denial that science constitutes, or could ever come to constitute, a single, unified project. The second is metaphysical, a thesis about how the world is. This thesis is an assertion of the extreme diversity of the contents of the world. There are countless kinds of things, I maintain, subject each to its own characteristic behavior and interactions. In addition, I propose a relation between these two theses: the second shows the inevitability of the first." John Dupré, *The Disorder of Things: Metaphysical Foundations of the Disunity of Knowledge* (Cambridge, MA: Harvard University Press, 1993), 1. It is worth noting that Kant could not have approved of the second thesis that Dupré puts forward: It is one thing to say that the world *may be* characterized by the extreme, irreducible diversity of its components (Kant), and another to say that the world *is* characterized by such a diversity (Dupré). What I call "critical naturalism" confines itself to the first thesis.

33. Julian S. Huxley, *Religion Without Revelation* (New York and London: Harper and Brothers, 1927), 137.

34. The debunking of scientistic fallacies distinguishes critical naturalism from milder forms of methodological naturalism as defined, for instance, by Robert T. Pennock, "Naturalism, Evidence, and Creationism: The Case of Phillip Johnson," *Biology and Philosophy* 11, no. 4 (1996): 543–559.

Chapter 4

1. "In the traditional view a person is free. He is autonomous in the sense that his behavior is uncaused. He can therefore be held responsible for what he does and justly punished if he offends. That view, together with its associated practices, must be re-examined when a scientific analysis reveals unsuspected controlling relations between behaviour and environment. . . . A scientific analysis naturally moves in the direction of clarifying all kinds of controlling relations. By questioning the control exercised by autonomous man and demonstrating the control exercised by the environment, a science of behaviour also seems to question dignity and worth." B. F. Skinner, *Beyond Freedom and Dignity* (London: Penguin Books, 1971), 25–26.

2. Thomas Lemke, *Biopolitics: An Advanced Introduction* (New York and London: New York University Press, 2011), 119.

3. Lemke, *Biopolitics*, 119–121.

4. With regard to Italian biopolitical theory, the reference works are Giorgio Agamben, *Homo Sacer: Sovereign Power and Bare Life* (Stanford, CA: Stanford University Press, 1998); Michael Hardt and Antonio Negri, *Empire: The New World Order* (Cambridge, MA: Harvard University Press, 2000); Roberto Esposito, *Bios: Biopolitics and Philosophy* (Minneapolis: University of Minnesota Press, 2008). On biopolitics, capitalism, and neoliberal economy, see Kaushik Sunder Rajan, *Biocapital: The Constitution of Postgenomic Life* (Durham, NC: Duke University Press, 2006); Kaushik Sunder Rajan, *Pharmocracy: Value, Politics, and Knowledge in Global Biomedicine* (Durham, NC: Duke University Press, 2017); Melinda Cooper, *Life as Surplus: Biotechnology and Capitalism in the Neoliberal Era* (Seattle: University of Washington Press, 2008). On the new forms of biosubjectivation and biosocialization, see Donna Haraway, *Simians, Cyborgs, and Women: The Reinvention of Nature* (London: Free Association Books, 1991); Paul Rabinow, *Essays on the Anthropology of Reason* (Princeton, NJ: Princeton University Press, 1996); Paul Rabinow, *Marking Time: On the Anthropology of the Contemporary* (Princeton, NJ: Princeton University Press, 2008); Paul Rabinow and Nikolas Rose, "Biopower Today," *Biosocieties* 1, no. 2 (2006): 195–217; Nikolas Rose, *The Politics of Life Itself: Biomedicine, Power, and Subjectivity in the Twenty-First Century* (Princeton, NJ: Princeton University Press, 2007); Rosi Braidotti, "Posthuman Affirmative Politics," in *Resisting Biopolitics: Philosophical, Political, and Performative Strategies*, ed. S. E. Wilmer and Audronė Žukauskaitė (London: Routledge, 2016), 30–56. On the new social imaginaries in biopolitical times, see Barbara Duden, *Disembodying Women: Perspectives on Pregnancy and the Unborn* (Cambridge, MA: Harvard University Press, 1993); Sarah Franklin, "Life Itself: Global Nature and the Genetic Imaginary," in *Global Nature, Global Culture*, ed. Sarah Franklin, Celia Lury, and Jackie Stacey (London: Sage Publications, 2000), 188–227.

5. Lemke, *Biopolitics*, 15–21.

6. See, for instance, Thomas C. Wiegele, *Biopolitics: Search for a More Human Political Science* (Boulder, CO: Westview Press, 1979); Thomas C. Wiegele, ed., *Biology and the Social Sciences: An Emerging Revolution* (Boulder, CO: Westview Press, 1982); Albert Somit and Steven A. Peterson, "Introduction: Main Currents in Biopolitics," *International Political Science Review* 8, no. 2

(1987): 107–110; Robert H. Blank and Samuel M. Hines, Jr., *Biology and Political Science* (London: Routledge, 2001); Albert Somit and Steven A. Peterson, eds., *Human Nature and Public Policy: An Evolutionary Approach* (London: Palgrave Macmillan, 2003); Albert Somit and Steven A. Peterson, *The Failure of Democratic Nation Building: Ideology Meets Evolution* (London: Palgrave Macmillan, 2005); Nancy Meyer-Emerick, "Public Administration and the Life Sciences," *Administration & Society* 38, no. 6 (2007): 689–708; John R. Alford and John R. Hibbing, "The New Empirical Biopolitics," *Annual Review of Political Science* 11 (2008): 183–203; Albert Somit, "What Hath Biopolitics Wrought?" *Politics and the Life Sciences* 30, no. 1 (2011): 95–97; Albert Somit and Steven A. Peterson, eds., *Biopolicy: The Life Sciences and Public Policy* (Bingley, UK: Emerald, 2012).

7. Michel Foucault, "The Subject and Power," in *Power: Essential Works of Foucault, 1954–1984*, vol. 3 (New York: New Press, 2003), 330.

8. "There are phenomena which cannot be analyzed and traced back to other phenomena. They are the ultimate given. The progress of scientific research may succeed in demonstrating that something previously considered as an ultimate given can be reduced to components. But there will always be some irreducible and unanalyzable phenomena, some ultimate given. . . . Action is the essence of his nature and existence, his means of preserving his life and raising himself above the level of animals and plants. However perishable and evanescent all human efforts may be, for man and for human science they are of primary importance." Ludwig von Mises, *Human Action: A Treatise on Economics* (San Francisco: Fox & Wilks, 1996), 17–19.

9. See, for instance, Tibor Scitovsky, *The Joyless Economy: The Psychology of Human Satisfaction* (Oxford: Oxford University Press, 1976).

10. "The problem is with the model being used by economists, a model that replaces homo sapiens with a fictional creature called *homo economicus*, which I like to call an Econ for short. Compared to this fictional world of Econs, Humans do a lot of misbehaving, and that means that economic models make a lot of bad predictions, predictions that can have much more serious consequences than upsetting a group of students. . . . To understand the consumption behavior of households, we clearly need to get back to studying Humans rather than Econs. Humans do not have the brains of Einstein (or Barro), nor do they have the self-control of an ascetic Buddhist monk. Rather, they have passions, faulty telescopes, treat various pots of wealth quite differently, and can be influenced by short-run returns in the stock market. We need a model of these kinds of Humans." Richard H. Thaler, *Misbehaving: The Making of Behavioral Economics* (New York: W. W. Norton, 2016), 16, 105.

11. See Herbert A. Simon, *Models of Bounded Rationality* (Cambridge, MA: MIT Press, 1997).

12. The same kind of assumption about the unreliable rationality of the single individual also lies behind the theory of "rational expectations" (developed by John A. Muth, Robert E. Lucas Jr., and others), which is based on the idea that individuals operating in the market may be wrong some of the time, as each of them has access to limited knowledge, but the population (i.e., the "markets") will nonetheless make optimal decisions *on average*. From this perspective, economists can hold firm to the principle that people behave in ways that maximize their utility while giving up the belief in *homo economicus*. "The concept of rational expectations asserts that outcomes do not differ systematically (i.e., regularly or predictably) from what people expected them to be. The concept is motivated by the same thinking that led Abraham Lincoln to assert, 'You can fool some of the people all of the time, and all of the people some of the time, but you cannot fool all of the people all of the time.' From the viewpoint of the rational expectations doctrine, Lincoln's statement gets things right. It does not deny that people often make forecasting errors, but it does suggest that errors will not persistently occur on one side or the other. Economists who believe in rational expectations base their belief on the standard economic assumption that people behave in ways that maximize their utility (their enjoyment of life) or profits." Thomas J. Sargent, "Rational Expectations," in *Fortune Encyclopedia of Economics*, ed. David R. Henderson (New York: Warner Books, 1993), 155–156.

13. "Conventional economic analyses of policy recognize only one measure of the value of public goods: the aggregate willingness of the public to pay for them. There are serious doubts about the coherence of this concept and the feasibility of measuring willingness to pay. A more fundamental question is whether willingness to pay should remain the only measure of value. The present analysis suggests that moment-based measures of the actual experience of consequences should be included in assessments of outcomes and as one of the criteria for the quality of decisions, both public and private." Daniel Kahneman, "Experienced Utility and Objective Happiness: A Moment-Based Approach," in *Choices, Values, and Frames*, ed. Daniel Kahneman and Amos Tversky (Cambridge: Cambridge University Press, 2000), 690.

14. As a critic has observed, the problem with this approach is that the "object of Kahneman and Tversky's inquiry is not, cannot be, the *individual* decision-maker. The object of their inquiry is the *population* of individuals from which their rather arbitrary sample of subjects was drawn. This is a limitation of these studies as contributions to psychology—but perhaps not to economics. (In fact, they seem to have led to further developments in economics rather than to advances in our understanding of the process of individual choice.)" John Staddon, *The New Behaviorism: Foundations of Behavioral Science* (New York: Routledge, 2021), 125.

15. Michel Foucault, "The Confession of the Flesh," in *Power/Knowledge. Selected Interviews and Other Writings 1972–1977* (New York: Pantheon Books 1980), 195.

16. Here, I have slightly modified the English translation of *faire mourir et laisser vivre* and *faire vivre et rejeter dans la mort* (as "to take life and let live" and "to foster life and disallow it to the point of death"), which can be found in Michel Foucault, *The History of Sexuality. Volume I: An Introduction* (New York: Pantheon Books, 1978), 138.

17. Michel Foucault, "La politique de la santé au XVIII siècle," in *Dits et écrits*, vol. 3 (Paris: Gallimard-Seuil, 1994), 731.

18. For more details, see, for instance, Rodolfo Saracci, "Introducing the History of Epidemiology," in *Teaching Epidemiology*, ed. Jorn Olsen (Oxford: Oxford University Press, 2001), 3–29; Alfredo Morabia, *A History of Epidemiological Methods and Concepts* (Basel, Switzerland: Birkhäuser, 2004); Kenneth J. Rothman, *Epidemiology: An Introduction* (Oxford: Oxford University Press, 2012). On Nazi epidemiology and public health policies, see Robert N. Proctor, *The Nazi War on Cancer* (Princeton, NJ: Princeton University Press, 2000).

19. On the importance of William Farr in the history of epidemiology, see, for instance, David E. Lilienfeld, "Celebration: William Farr (1807–1883)—An Appreciation on the 200th Anniversary of His Birth," *International Journal of Epidemiology* 36, no. 5 (2007): 985–987.

20. This presentation of the epistemological foundations of epidemiology is oversimplified. For a short introduction to the topic, see Mark Parascandola and D. L. Weed, "Causation in Epidemiology," *Journal of Epidemiology and Community Health* 55, no. 12 (2001): 905–912; Mark Parascandola, "The Epidemiological Transition and Changing Concepts of Causation and Causal Inference," *Revue d'Histoire des Sciences* 64, no. 2 (2011): 243–262.

21. As Doll and Hill remark at the end of their pathbreaking article, "As to nature of the carcinogen we have no evidence. . . . From consideration of the smoking histories by the patients without cancer of the lung a tentative estimate was made of the number of people who smoked different amounts of tobacco in Greater London, and hence the relative risks of developing the disease among different grades of smokers were calculated. The figure obtained are admittedly speculative, but suggest that, above the age of 45, the risk of developing the disease increases in simple proportion with the amount smoked, and that it may be approximately 50 times as great among those who smoke 25 or more cigarettes a day as among non-smokers." Richard Doll and A. Bradford Hill, "Smoking and Carcinoma of Lung: Preliminary Report," *British Medical Journal* 2, no. 4682 (1950): 747.

22. It is perhaps worth noting that the vast sociological literature on the risk society most often overlooks the role played by the new epidemiology in the framing and rise of a "new modernity." See, for instance, Ulrich Beck, *Risk Society: Towards a New Modernity* (London: Sage, 1992); Barbara Adam, Ulrich Beck, and Joost van Loon, eds., *The Risk Society and Beyond* (London: Sage, 2000).

23. "It is important to appreciate that in epidemiological usage, 'risk' is a purely statistical concept." Alex Broadbent, *Philosophy of Epidemiology* (London: Palgrave Macmillan, 2013), 22.

24. On the government of risks see also Robert Castel, "From Dangerousness to Risk," in *The Foucault Effect: Studies in Governmentality*, ed. Graham Burchell, Colin Gordon, and Peter Miller (Chicago: University of Chicago Press, 1991), 281–298; Robert Castel, *La Gestion des riques* (Paris: Minuit, 2011); François Ewald, "Two Infinities of Risk," in *The Politics of Everyday Fear*, ed. Brian Massumi (Minneapolis: University of Minnesota Press, 1993), 221–228.

25. On the concept of "governmentality" as understood by Foucault and governmentality studies after him, see Ulrich Bröckling, Susanne Krasmann, and Thomas Lemke, "From Foucault's Lectures at the Collège de France to Studies of Governmentality: An Introduction," in *Governmentality: Current Issues and Future Challenges*, ed. Ulrich Bröckling, Susanne Krasmann, and Thomas Lemke (New York: Routledge, 2011), 1–33.

26. For a critical examination of the role played by evaluations today, see Jacques-Alain Miller and Jean-Claude Milner, *Évaluation. Entretiens sur une machine d'imposture* (Paris: Agalma, 2004).

27. As Alex Broadbent points out, population thinking is among the most striking features of epidemiology: "The fifth striking feature is certainly not unique: it is the centrality of population thinking. This term is familiar from other contexts, notably the philosophy of biology. In epidemiology, its importance consists in the idea that populations may be thought of as bearing health-related properties. This is sometimes counterintuitive, since it is individuals who suffer diseases. But measuring the level of a disease in a population is central to epidemiology, and it requires thinking of populations as entities which can bear properties. There are a number of philosophically interesting problems here. Some are general: for example, we can ask whether population thinking is merely instrumental or whether populations really are property bearers. Some are more specific to epidemiology: for example, how population properties relate to individual ones, what inferences from one to the other are licensed, and even, in some contexts, what inferences from population to individual are ethically or legally warranted." Broadbent, *Philosophy of Epidemiology*, 7.

28. Noam Chomsky and Michel Foucault, *On Human Nature* (New York: New Press, 2006), 6.

29. See Michel Foucault, *The Order of Things: An Archaeology of the Human Sciences* (New York: Pantheon Books, 1970).

30. Prescriptions of antidepressants often follow this logic of scientific discovery: "One of the most popular theories posits that a chemical imbalance in the brain, specifically deficient amounts of the neurochemical serotonin, causes depression. It follows, then, that drug treatments targeting serotonin, the selective serotonin reuptake inhibitors (SSRIs), which presumably correct this chemical imbalance, are the appropriate response to depressive disorders. This theory is relentlessly promoted in many ways: pharmaceutical advertisements emphasize how correctable chemical imbalances cause depressive disorders, public service messages stress that depression stems from flaws in brain chemistry rather than in character, and mental health advocacy groups advance the message that depression is a physical, brain-based illness, just like diabetes or asthma. These ubiquitous messages have led to the widespread impression that research has actually shown that chemical deficiencies are the cause of depressive disorders and that drugs work because they correct these impairments in the neurotransmission system. Therefore, it might seem that one way to separate depressive disorders from normal sadness would be to examine levels of serotonin in the brain. The chemical deficiency theory of depression originated with psychiatrist Joseph Schildkraut's hypothesis, published in 1965, that low levels of amines were associated with the development of depressive disorders. . . . The major source of evidence for the chemical deficiency hypothesis stems from the success of drug treatments, which raise levels of amines, in alleviating depressive symptoms. Schildkraut himself recognized that 'even if the drugs are effective in treating the disorders, that does not necessarily imply that their mode of action involves correction of the underlying abnormality.' Nevertheless, many subsequent arguments relating to serotonin rely on the premise that if enhancing its transmission improves depression, then a deficiency in the serotonin system may be responsible for the initial emergence of the condition." Allan V. Horwitz and Jerome C. Wakefield, *The Loss of Sadness: How Psychiatry Transformed Normal Sorrow into Depressive Disorder* (Oxford: Oxford University Press, 2007), 168–169.

31. As for the definition of adaptive behavior, it varies slightly depending on the type of therapy. For Albert Ellis, one of the founding figures of the cognitive-behavioral approach, adaptive behavior is achieved when patients show "self-interest, social interest, self-direction, tolerance of self and others, flexibility, acceptance of uncertainty, commitment to vital interests, self-acceptance, scientific thinking, and a nonutopian perspective on life." Keith S. Dobson and David J. A. Dozois, "Historical and Philosophical Bases of the Cognitive-Behavioral Therapies," in *Handbook of Cognitive-Behavioral Therapies*, ed. Keith S. Dobson (London: Guilford Press, 2010), 13.

32. See, for instance, Oliver Burkeman, "Therapy Wars: The Revenge of Freud," *The Guardian*, January 7, 2016.

33. See Adam M. Volungis, *Cognitive-Behavioral Therapy: Theory into Practice* (Lanham, MD: Rowman & Littlefield, 2019), 311–320.

34. "Classical conditioning, or respondent conditioning, is derived from Pavlov's famous dog experiment, in which he trained dogs to salivate when a tuning fork was presented in the absence of food. Pavlovian conditioning theory suggests that anxiety disorders are classically conditioned when pathological fear (e.g., fear of flying) is associated with an objectively innocuous stimulus (e.g., an airplane), which has been learned in association with a feared stimulus (e.g., death by plane crash). The fear structure is created and maintained when a fear response, or conditioned response (CR; e.g., fear of flying) to a conditioned stimulus (CS; e.g., an airplane) occurs in the absence of the unconditioned stimulus (US; death by plane crash). This erroneous association is what causes a specific phobia to develop, specifically, in this example, aerophobia. This theory of classical conditioning is one of the building blocks of behaviorism, and many novel interventions have been developed as a result. One of the most widely used and efficacious treatments for anxiety disorders is exposure therapy, which utilizes the idea of Pavlovian extinction." Michelle L. Davis, Sara M. Witcraft, Scarlett O. Baird, and Jasper A. J. Smits, "Learning Principles in Cognitive Behavioral Therapy," in *The Science of Cognitive Behavioral Therapy*, ed. Stefan G. Hofman and Gordon J.G. Asmundson (London: Academic Press, 2017), 52.

35. "Reinforcement is any stimulus or event that increases, or strengthens, the frequency of a behavior. Positive reinforcement occurs when the addition of a positive stimulus or event strengthens a behavior and makes it more likely to occur in the future. Negative reinforcement occurs when a behavior is strengthened by the removal of an aversive event or stimulus. Conversely, punishment involves a stimulus or event that will decrease, or weaken, the likelihood of the behavior in the future. Positive punishment occurs when an aversive consequence is introduced to reduce a behavior. Negative punishment occurs when a positive stimuli or event is removed in order to decrease the frequency of a behavior." Davis et al., "Learning Principles in Cognitive Behavioral Therapy," 57.

36. The idea of exposure comes directly from the first behaviorists (including Ivan Pavlov and John B. Watson), who conducted their research in the early twentieth century: "The object of exposure is to present the feared object or situation to the child (either *in vivo* or *in vitro*) in a way that fosters new learning. A variety of behavioral theories have been proposed to support the notion that presentation of a conditioned stimulus (e.g., feared object) in the absence of the unconditioned stimulus (e.g., feared outcome) should eliminate the conditioned response. . . . In other words, presenting the signal for danger (e.g., dog) in the absence of the actual danger (e.g., bite) should weaken their association and hence reduce the fear response to the signal (i.e., dog)." Bruce F. Chorpita, *Modular Cognitive-Behavioral Therapy for Childhood Anxiety Disorders* (New York: Guilford Press, 2007), 53–54.

37. The theory of modeling was developed by Albert Bandura in the 1960s: "Modeling involves the child's observation of another person interacting successfully with a feared stimulus. For example, a therapist might pet a dog to demonstrate to a child that the dog is safe and need not be feared. There are many variations of this basic technique, depending primarily on the nature of the model and the type of behaviors expected of the child." Chorpita, *Modular Cognitive-Behavioral Therapy*, 45.

38. "Cognitive restructuring teaches clients how to elect to use experience as evidence to update or restructure unhelpful thoughts into more rational thoughts." Davis et al., "Learning Principles in Cognitive Behavioral Therapy," 61.

39. On this new "grammar" of power, see also Davide Tarizzo, *Political Grammars: The Unconscious Foundations of Modern Democracy* (Stanford, CA: Stanford University Press), 163–199.

40. Karl Pearson, *The Grammar of Science* (London: Adam and Charles Black, 1911), 12.

Chapter 5

1. "A normalizing society is the historical outcome of a technology of power centered on life." Michel Foucault, *The History of Sexuality. Volume I: An Introduction* (New York: Pantheon Books, 1978), 144.

2. Gilles Deleuze, "Postscript on the Societies of Control," *October* 59 (1992): 3–7.

3. Nikolas Rose, *The Politics of Life Itself: Biomedicine, Power, and Subjectivity in the Twenty-First Century* (Princeton, NJ: Princeton University Press, 2007), 27.

4. Nikolas Rose and Filippa Lentzos, "Making Us Resilient: Responsible Citizens for Uncertain Times," in *Competing Responsibilities: The Ethics and Politics of Contemporary Life*, ed. Susanna Trnka and Catherine Trundle (Durham, NC: Duke University Press, 2017), 27–48.

5. Deleuze, "Postscript on the Societies of Control," 3.

6. Foucault, *The History of Sexuality. Volume I*, 89.

7. Antonio Negri and Michael Hardt, *Empire: The New World Order* (Cambridge, MA: Harvard University Press, 2000), 23.

8. For a more accurate comparison between Foucault's analyses and Deleuze's view, see, for instance, Maurizio Lazzarato, "The Concepts of Life and the Living in the Societies of Control," in *Deleuze and the Social*, ed. Martin Fuglsang and Bert Meier Sorensen (Edinburgh: Edinburgh University Press, 2006), 171–190.

9. Here, I confine myself to summarizing some theses discussed in Foucault, *The History of Sexuality. Volume I*. But see also Michel Foucault, *Discipline and Punish: The Birth of the Prison* (New York: Random House, 1995); Michel Foucault, *Security, Territory, Population: Lectures at the Collège de France, 1977–1978* (London: Palgrave Macmillan, 2007).

10. Foucault, *Security, Territory, Population*, 85.

11. See, for instance, Georges Canguilhem, *The Normal and the Pathological*, with an introduction by Michel Foucault (New York: Zone Books, 1991).

12. Michel Foucault, "What Is Enlightenment?" in *The Foucault Reader*, ed. Paul Rabinow (New York: Pantheon Books, 1984), 50.

13. See, for instance, Michel Foucault, "Lives of Infamous Men," in *Power: Essential Works of Foucault*, vol. 3 (New York: New Press, 2003), 157–175.

14. Michel Foucault, *The Birth of Biopolitics: Lectures at the Collège de France, 1978–1979* (New York: Picador, 2010), 67.

15. Foucault, *The Birth of Biopolitics*, 67.

16. Foucault, *The Birth of Biopolitics*, 67.

17. See, for instance, Pierre Legendre, *Le crime du caporal Lortie: Traité sur le Père* (Paris: Fayard, 1989); Pierre Legendre, *La 901e conclusion: Étude sur le théâtre de la Raison* (Paris: Fayard, 1998); Jean-Claude Milner, *Les penchants criminels de l'Europe démocratique* (Paris: Verdier, 2003); Jean-Claude Milner, *Clartés de tout* (Paris: Verdier, 2011); Jean-Claude Poizat, "Entretien avec Jean-Claude Milner," *Le Philosophoire* 1, no. 43 (2015): 11–55.

18. Deleuze, "Postscript on the Societies of Control," 5.

19. For more details on the history of the Normal curve, see, for instance, Karl Pearson, "Historical Note on the Origin of the Normal Curve of Errors," *Biometrika* 16, no. 3–4 (1924): 402–404; Helen M. Walker, "Bi-centenary of the Normal Curve," *Journal of the American Statistical Association* 29, no. 185 (1934): 72–75.

20. Richard J. Herrnstein and Charles Murray, *The Bell Curve: Intelligence and Class Structure in American Life* (New York: Simon & Schuster, 1994).

21. "These methods, which made possible the meticulous control of the operations of the body, which assured the constant subjection of its forces and imposed upon them a relation of docility-utility, might be called 'disciplines'." Foucault, *Discipline and Punish*, 137.

22. Lionel Robbins, *An Essay on the Nature and Significance of Economic Science* (London: Macmillan, 1932), 27. Hilary Putnam, among others, has raised doubts about Robbins's moral neutralization of economic science, which nonetheless remains paradigmatic. See Hilary Putnam, *The Collapse of the Fact/Value Dichotomy and Other Essays* (Cambridge, MA: Harvard University Press, 2002).

23. Stephen Jay Gould, *The Structure of Evolutionary Theory* (Cambridge, MA: Harvard University Press, 2002), 230.

24. "In the societies of control, on the other hand, what is important is no longer either a signature or a number, but a code: the code is a *password*, while on the other hand the disciplinary societies are regulated by *watchwords* (as much from the point of view of integration as from that of resistance). The numerical language of control is made of codes that mark access to information, or reject it. We no longer find ourselves dealing with the mass/individual pair. Individuals have become '*dividuals*'." Deleuze, "Postscript on the Societies of Control," 5.

25. For more details on Quetelet and *l'homme moyen*, see Kevin Donnelly, *Adolphe Quetelet, Social Physics and The Average Men of Science, 1796–1874* (Pittsburgh: University of Pittsburgh Press, 2016).

26. "The disciplinary societies have two poles: the signature that designates the *individual*, and the number or administrative numeration that indicates his or her position within a *mass*. This is because the disciplines never saw any incompatibility between these two, and because at the same time power individualizes and masses together, that is, constitutes those over whom it exercises power into a body and molds the individuality of each member of that body." Deleuze, "Postscript on the Societies of Control," 5.

27. See, for instance, Michel Foucault, *"Society Must Be Defended": Lectures at the Collège de France, 1975–1976* (New York: Picador, 2003), 249–252.

28. For more details on *homo nationalis*, see, for instance, Étienne Balibar, *Nous, citoyens d'Europe? Les frontières, l'État, le peuple* (Paris: La Découverte, 2001); Étienne Balibar, *Politics and the Other Scene* (London: Verso, 2002).

29. Michel Foucault, "The Subject and Power," in *Power: Essential Works of Foucault*, vol. 3, 331.

30. Foucault, "The Subject and Power," 331.

31. David L. Rosenhan, "On Being Sane In Insane Places," *Science* 179 (1973): 257.

32. Rosenhan, "On Being Sane In Insane Places," 253–254.

33. Rosenhan, "On Being Sane In Insane Places," 250–251.

34. Rosenhan, "On Being Sane In Insane Places," 254.

35. The veracity of the Rosenhan experiment has been called into doubt by Susannah Cahalan, *The Great Pretender: The Undercover Mission That Changed Our Understanding of Madness* (New York: Grand Central, 2019).

36. See Robert L. Spitzer, "On Pseudoscience in Science, Logic in Remission, and Psychiatric Diagnosis: A Critique of Rosenhan's 'On Being Sane in Insane Places'," *Journal of Abnormal*

Psychology 84, no. 5 (1975): 442–452; Robert L. Spitzer, "More on Pseudoscience in Science and the Case for Psychiatric Diagnosis," *Archive of General Psychiatry* 33 (1976): 459–470.

37. The key distinction between reliability and validity bears emphasizing: "Although efforts to produce more reliable classifications operate entirely on the empirical field, at the level of operational criteria, observations, measures, and empirical indicants, validity always involves a two-level process—the first being conceptual and abstract, and the second being empirical and observed. Validity concerns the correspondence between our empirical indicants, our diagnostic criteria and measurement methods, and the constructs and ideas onto which we hope they will map. Reliability is entirely an empirical process, whereas validity forces us to grapple with our fundamental concepts of mental disorder." Stephen L. Buka and Stephen E. Gilman, "Psychopathology and the Life Course," in *Defining Psychopathology in the 21st Century: DSM-V and Beyond*, ed. John E. Helzer and James J. Hudziak (Washington, DC: American Psychiatric Publishing, 2002), 130. On this issue, see also Nick Haslam, "Reliability, Validity, and the Mixed Blessings of Operationalism," in *Oxford Handbook of Philosophy and Psychiatry*, ed. K. W. M. Fulford, Martin Davies, Richard G. T. Gipps, et al. (Oxford: Oxford University Press, 2013), 987–1002.

38. Roger K. Blashfield and Danny R. Burgess, *Classification Provides an Essential Basis for Organizing Mental Disorders*, in *The Great Ideas of Clinical Science*, ed. Scott O. Lilienfeld and William T. O'Donohue (New York: Routledge, 2007), 98–99.

39. Rosenhan, "On Being Sane In Insane Places," 254.

40. As Skinner explained in his 1945 paper on operationism, "The operational attitude, in spite of its shortcomings, is a good thing in any science but especially in psychology because of the presence there of a vast vocabulary of ancient and non-scientific origin. It is not surprising that the broad empirical movement in the philosophy of science, which Stevens has shown to be the background of operationism, should have had a vigorous and early representation in the field of psychology—namely, behaviorism. In spite of the differences which Stevens claims to find, behaviorism has been (at least to most behaviorists) nothing more than a thoroughgoing operational analysis of traditional mentalistic concepts." B. F. Skinner, "The Operational Analysis of Psychological Terms," *Psychological Review* 52 (1945): 271. See also Stanley S. Stevens, "The Operational Definition of Psychological Concepts," *Psychological Review* 42 (1935): 517–527; Emilio Ribes-Iñesta, "What Is Defined in Operational Definitions? The Case of Operant Psychology," *Behavior and Philosophy* 31 (2003): 111–126.

41. The concept of operational definition was introduced by Percy W. Bridgman, *The Logic of Modern Physics* (New York: Macmillan, 1927). For a philosophical discussion of the concept, see Carl Gustav Hempel, *Aspects of Scientific Explanation* (New York: Free Press, 1965), 123–133; Carl Gustav Hempel, *Philosophy of Natural Science* (Upper Saddle River, NJ: Prentice Hall, 1966), 278–290. For more details on Bridgman and Skinner, see Harold J. Allen, "P. W. Bridgman and B. F. Skinner on Private Experience," *Behaviorism* 8, no. 1 (1980): 15–29. On operationism and experimental psychology in the US before DSM-III, see Uljana Feest, "Operationism in Psychology: What the Debate Is About, What the Debate Should Be About," *Journal of the History of Behavioral Sciences* 41, no. 2 (2005): 131–145.

42. "One significant advance of DSM-III over prior editions of the DSM was its provision of operationalized diagnostic criteria that defined the various mental disorders in terms of their symptomatic presentations, rather than in terms of unproven theoretically based assumptions about their etiology." Anthony F. Lehman, George S. Alexopoulos, and Dilip Jeste, "Mental Disorders and

Disability," in *A Research Agenda for DSM-V*, ed. David J. Kupfer, Michael B. First, and Darrel A. Regier (Washington, DC: American Psychiatric Association, 2002), 202.

43. "The political controversy over the task force's deletion of 'neurosis' as a diagnosis most clearly illustrated the political aspects of the DSM-III's crafting. It became the most hotly contested and divisive struggle in the DSM-III's ratification. Neurosis was both a synthesizing rationale and a fundamental concept for Freudian psychoanalysts. The term, as used by advocates of psychotherapy, described the underlying process of internal psychological conflict present in virtually all individuals, which resulted in symptoms that served to unconsciously control anxiety. Everyone, then, existed on a continuum of minimal to severe neurosis. Neurosis had evolved into a staple diagnostic term that many psychiatrists used. Freudian psychoanalysts were especially fond of it as justification for their services—primarily different forms of long-term, individual talk therapy—when no other more specific diagnosis seemed to fit an individual's mental symptoms. Spitzer and his colleagues had intended to delete neurosis from the outset of their efforts. In their opinion, neurosis had no empirical basis in fact." Rick Mayes and Allan V. Horwitz, "DSM-III and the Revolution in the Classification of Mental Illness," *Journal of the History of the Behavioral Sciences* 41, no. 3 (2005): 261–262.

44. American Psychiatric Association, *Diagnostic and Statistical Manual of Mental Disorders*, 3rd ed. (Washington, DC: American Psychiatric Association, 1980), 6.

45. Allan V. Horwitz, *DSM: A History of Psychiatry's Bible* (Baltimore: John Hopkins University Press, 2021), 66.

46. Karl Menninger, *The Vital Balance* (New York: Viking Press, 1963), 325.

47. Paul E. Meehl, "Diagnostic Taxa as Open Concepts: Metatheoretical and Statistical Questions about Reliability and Construct Validity in the Grand Strategy of Nosological Revision," in *Contemporary Directions in Psychopathology: Scientific Foundations of DSM-V and ICD-11*, ed. Theodore Millon, Robert F. Krueger, and Erik Simonsen (New York: Guilford Press, 2010), 178–179; originally published in *Contemporary Directions in Psychopathology*, ed. Theodore Millon and Gerald L. Klerman (New York: Guilford Press, 1986), 215–231.

48. "The mistaken and intellectually indefensible view that the DSM constructs represent the truth is indeed bad news. Not only can such a misguided view get in the way of progress; it is also grounded in a view of entities and operational definitions that has long been abandoned by informed philosophers of science and psychological science researchers. What is meant here? If someone believes that the *meaning* of the conditions in DSM is defined by the list of signs and symptoms (or DSM-defined diagnostic rules), then one is implicitly subscribing to an 'operationism' (so-called operational definitions) that has long been discounted in philosophy and psychology. Conditions (or diseases) in traditional organic medicine are defined not merely by signs and symptoms; rather, they represent (implicitly) information regarding etiology, pathophysiology, and so on. If information about etiology and pathophysiology is absent from such constructs in organic medicine, then, at a minimum, the construct in question is viewed as in need of further research. In short, the constructs defined explicitly in DSM do not represent the truth, so to speak, and they do not represent the intellectual end of the line. For readers who might think that this is self-evident stuff, I might only remind them of the frequent precursor to the modal question that follows many conference talks or colloquia—namely, 'According to DSM-IV, yadda, yadda, yadda.' I think that a routine rejoinder to such a preface to commentary after talks should be 'Just what do you believe about the constructs in DSM?'" Mark F. Lenzenweger,

"Contemplations on Meehl (1986): The Territory, Paul's Map, and Our Progress in Psychopathology Classification (or, the Challenge of Keeping up with a Beacon 30 Years Ahead of the Field)," in *Contemporary Directions in Psychopathology: Scientific Foundations of DSM-V and ICD-11*, ed. Theodore Millon, Robert F. Krueger, and Erik Simonsen (New York: Guilford Press, 2010), 193.

49. For more details on "DSM-5's failed revolution," see Horwitz, *DSM: A History of Psychiatry's Bible*, 116–143.

50. Dennis G. Fisher, Robert Malow, Rhonda Rosenberg, Grace L. Reynolds, Nisha Farrell, and Adi Jaffe, "Recreational Viagra Use and Sexual Risk among Drug Abusing Men," *American Journal of Infectious Diseases* 2, no. 2 (2006): 107–114.

51. Berhanemeskel Weldegerima Atsbeha, Beza Tefera Kebede, Biruktawit Shewafera Birhanu, Dawit Kumilachew Yimenu, Wudneh Simegn Belay, and Chilot Abiyu Demeke, "The Weekend Drug; Recreational Use of Sildenafil Citrate and Concomitant Factors: A Cross-Sectional Study," *Frontiers in Medicine* 8 (2021): doi.org/10.3389/fmed.2021.665247.

52. Tazin Karim, *Medicating Our Children: The Moral Influence of Adderall on Education, Parenting, and Treatment*, somatosphere.net/2012/10/meducating-our-children-the-moral-influence-of-adderall-on-education-parenting-and-treatment.html.

53. Bill Readings, *The University in Ruins* (Cambridge, MA: Harvard University Press, 1997).

54. Alan Schwarz, "Attention Disorder or Not, Pills to Help in School," *New York Times*, October 9, 2012. As some transhumanist philosophers have emphasized, "Drugs such as metylphenidate (Ritalin) are already used to treat ADHD and occasionally also for enhancement purposes." Nick Bostrom and Anders Sandberg, "The Wisdom of Nature: An Evolutionary Heuristic for Human Enhancement," in *Human Enhancement*, ed. Julian Savulescu and Nick Bostrom (Oxford: Oxford University Press, 2009), 390. For a brief overview of ethical issues related to neurocognitive enhancement, see Martha J. Farah, Judy Illes, Robert Cook-Deegan, et al., "Neurocognitive Enhancement: What Can We Do and What Should We Do?" *Nature Reviews Neuroscience* 5, no. 5, 421–425.

55. Gary S. Becker, *Human Capital: A Theoretical and Empirical Analysis, with Special Reference to Education* (Chicago: University of Chicago Press, 1993).

56. www.who.int/news-room/fact-sheets/detail/depression.

57. For the definition of depression as "a common mental disorder," see www.who.int/health-topics/depression#tab=tab_1.

58. On the history of depression, see also Alain Ehrenberg, *The Weariness of the Self: Diagnosing the History of Depression in the Contemporary Age* (Montreal and Kingston, Jamaica: McGill-Queen's University Press, 2010).

59. Horwitz, *DSM: A History of Psychiatry's Bible*, 70.

60. In 2013, DSM-5 has further enlarged the spectrum of depressive disorders by canceling the so-called Bereavement Exclusion, which excluded conditions that arose after the death of a beloved person.

61. Also, the elusiveness of the concept of "disability" can be traced to DSM-III, where it served to define the key concept of "disorder." See Horwitz, *DSM: A History of Psychiatry's Bible*, 62–63.

62. I found this passage on the WHO website more than ten years ago: www.who.int/topics/disabilities/en. Today, this webpage no longer exists.

63. World Health Organization, "Disability," www.who.int/health-topics/disability#tab=tab_1.

64. Foucault, *The Birth of Biopolitics*, 259–260.

65. In the fall of 1982, Foucault was still standing by his concept of normalization: "*Question*: You conclude *Discipline and Punish* by saying that it will 'serve as a background for various studies of normalization and the power of knowledge in modern society.' What is the relationship of normalization and the concept of man as the center of knowledge? *Answer*: Through these different practices—psychological, medical, penitential, educational—a certain idea or model of humanity was developed, and now this idea of man has become normative, self-evident, and is supposed to be universal." Michel Foucault, *Technologies of the Self*, ed. Luther H. Martin, Huck Gutman, and Patrick Hutton (Amherst: University of Massachusetts Press, 1988), 14–15.

Chapter 6

1. David Cayley, *Ivan Illich in Conversation* (Toronto: Anansi Press, 1992), 276.

2. Cayley, *Ivan Illich in Conversation*, 278.

3. The notion of a "government of the living" is taken from Michel Foucault, who famously claimed that "modern man is an animal whose politics places his existence as a living being in question." Michel Foucault, *The History of Sexuality. Volume I: An Introduction* (New York: Pantheon Books, 1978), 143.

4. John Maynard Smith, *The Theory of Evolution* (Harmondsworth, UK: Penguin, 1975), 96–97.

5. Stephen Jay Gould and Richard Lewontin, "The Spandrels of San Marco and the Panglossian Paradigm: A Critique of the Adaptationist Programme," *Proceedings of the Royal Society B* 205, no. 1161 (1979): 581.

6. "We criticize this approach and attempt to reassert a competing notion (long popular in continental Europe) that organisms must be analysed as integrated wholes, with *Baupläne* so constrained by phyletic heritage, pathways of development and general architecture that the constraints themselves become more interesting and more important in delimiting pathways of change than the selective force that may mediate change when it occurs." Gould and Lewontin, "The Spandrels of San Marco and the Panglossian Paradigm," 581.

7. John von Neumann, "The General and Logical Theory of Automata," in *Collected Works. Volume V: Design of Computers, Theory of Automata and Numerical Analysis* (Oxford, UK: Pergamon Press, 1963), 289.

8. von Neumann, "The General and Logical Theory of Automata," 289.

9. von Neumann, "The General and Logical Theory of Automata," 289–290.

10. von Neumann, "The General and Logical Theory of Automata," 290. See also John von Neumann, *Theory of Self-Reproducing Automata* (Urbana: University of Illinois Press, 1966).

11. B. F. Skinner, *The Behavior of Organisms: An Experimental Analysis* (New York: Appleton-Century, 1938), 6.

12. This definition is often quoted to illustrate the "strong" interpretation of Artificial Life research, according to which software can "synthesize" life, and not only "simulate" it, as the proponents of the "weak" interpretation maintain.

13. For more details on this topic, see, for instance, Ernst Mayr, "The Objects of Selection," *Proceedings of the National Academy of Sciences of the USA* 94 (1997): 2091–2094.

14. See Charles J. Lumsden and Edward O. Wilson, *Genes, Mind, and Culture: The Coevolutionary Process* (Cambridge, MA: Harvard University Press, 1981).

15. "Lumsden and Wilson neglect the dynamic properties of cultural inheritance when they come to consider its adaptive advantages. This leads to an overly simplified view of culture; one which seems to us cause Lumsden and Wilson to overestimate the role of genetic variation in explaining human differences." Robert Boyd and Peter J. Richerson, "Why Is Culture Adaptive?" *Quarterly Review of Biology* 58 (1983): 214. Elsewhere, Boyd and Richerson remark that, from a broader perspective, some "substantial modification of neo-Darwinian theory" must be introduced before the issue of cultural evolution can be addressed: "Neo-Darwinian theory, in its simplest form, holds that the phenotype of any individual organism is completely determined by its genotype; if one knew the genotype in complete detail, one could, in theory at least, predict the individual's phenotype. A more sophisticated view takes into account the interactive nature of genotype and the environment under the rubric of phenotypic plasticity. . . . Some social biologists appear to treat the capacity for culture as a biological adaptation, reducible in the ultimate analysis to the action of natural selection on genotype . . . The 'all genes' hypothesis cannot, for the reasons outlined above, accommodate the obvious fact that humans inherit at least some of their behavior culturally." Peter J. Richerson and Robert Boyd, "A Dual Inheritance Model of the Human Evolutionary Process I: Basic Postulates and a Simple Model," *Journal of Social and Biological Structures* 1, no. 2 (1978): 128.

16. Edward O. Wilson and B. F. Skinner, *A Dialogue between Sociobiology and Radical Behaviorism*, ed. Paul Naour (New York: Springer, 2009), 62.

17. Wilson and Skinner, *A Dialogue between Sociobiology and Radical Behaviorism*, 69, 67, 63.

18. Wilson and Skinner, *A Dialogue between Sociobiology and Radical Behaviorism*, 68. I have replaced "Al Knoll" (in the edited transcript) with "Al Newell," who actually gave the William James lectures at Harvard in 1987.

19. Christopher G. Langton, "Artificial Life," in *Artificial Life*, ed. Christopher G. Langton (Redwood City, CA: Addison-Wesley, 1989), 1–3.

20. B. F. Skinner, "Selection by Consequences," in *Upon Further Reflection* (Englewood Cliffs, NJ: Prentice-Hall, 1987), 51.

21. "Stimuli are always acting upon an organism, but their functional connection with operant behavior is not like that in the reflex. Operant behavior, in short, is *emitted*, rather than *elicited*." B. F. Skinner, *Science and Human Behavior* (New York: Free Press, 1965), 107.

22. Skinner himself sometimes seems aware of the problem: "The process of operant conditioning presumably evolved when those organisms which were more sensitively affected by the consequences of their behaviour were better able to adjust to the environment and survive. Only fairly immediate consequences could be effective. One reason for this has to do with 'final causes.' Behaviour cannot really be affected by anything which follows it, but if a 'consequence' is immediate, it may overlap the behaviour." B. F. Skinner, *Beyond Freedom and Dignity* (London: Penguin Books, 1971), 119.

23. Alexander Rosenberg objects to Skinner's theory along similar lines: "Roughly, the law of effect claims that once emitted, for whatever cause, a bit of behavior is likely to be repeated (or otherwise strengthened) if it is followed by some sort of benefit—the reinforcement—to the organism. . . . Skinner first exploited the law of effect in the explanation of the relatively complex behavior of rats and pigeons. By providing them with food pellets after the emission of some behavior like bar pressing or key pecking, he was able to get them to emit this behavior in a predictable fashion. . . . One thing stands out about this theory. Despite Skinner's claims, it is a thoroughly teleological one. It explains a bit of behavior in terms of apparently subsequent reinforcement. For example, people speak grammatically because they are reinforced after doing so (by smiling responses) and punished after failing to do so (by grimaces and corrections). Here we have a claim just like the biological assertion that the heart beats because its beating is followed by the blood's circulating. . . . The behavior the rat learns is thus defined by reference to its goal. But first, this definition is improperly teleological; second, the food is sometimes not there to be found, yet the behavior is the same. The obvious solution to this puzzle is to say that what the rat has learned isn't a set of movements with some end, goal, or purpose. What it has learned is where the food usually is. But that means it has learned something about the world—something that can be expressed in a proposition ('the food is at the end of the left alley'). To have been learned by the rat, this proposition must be *represented* in some state or other of the rat. We may not want to call this representation in the rat's brain a state of belief about where the food is. But it is still an intentional state of some kind." Alexander Rosenberg, *The Philosophy of Social Science* (Boulder, CO: Westview Press, 2016), 83, 84, 87.

24. "The main thing on which I feel I differ from you is that sociobiology leaps a little too cavalierly from socio- to bio-. Sociobiology leaves me out. I'm in the middle." Wilson and Skinner, *A Dialogue between Sociobiology and Radical Behaviorism*, 62.

25. As Charles Taylor remarked in his classic study on behaviorism in 1964, in which he did not pay much attention to Skinner's theory, a teleological explanation of behavior is also adopted by those trends of behaviorism that regard pleasure, rather than survival, as the motor of behavioral adaptation: "We seem to be adopting a teleological type of explanation in saying that whatever leads to pleasure will tend to occur; that is, we are explaining behaviour by what it tends to bring about. Thus in trying to avoid a teleological form of explanation of adaptation in terms of the goal of survival we will simply have introduced another teleological explanation in terms of the goal of pleasure. In other words, once one has posited a connexion between a certain action and pleasure, the problem still remains of explaining how this association will affect behaviour. This is the problem of 'getting to' the response. And the only way to solve this seems to be to invoke a teleological principle." Charles Taylor, *The Explanation of Behaviour* (New York: Routledge, 2021), 126.

26. B. F. Skinner, *Contingencies of Reinforcement: A Theoretical Analysis* (New York: Meredith Corporation, 1969), 297.

27. Skinner, *Contingencies of Reinforcement*, 294.

28. Skinner, *Beyond Freedom and Dignity*, 104.

29. As John Staddon emphasizes in his analysis of Skinner's philosophy, "Skinner, like many scientists who dabble in philosophy, seemed to believe that what *ought to be* can be inferred from *what is*—that science and reason can guide us in all things. But eighteenth-century philosopher David Hume showed this to be a fallacy. . . . Skinner, like contemporary *scientific naturalists*, does not accept Hume's conclusion. His Harvard colleague, the great biologist E. O. Wilson, for example, wrote that *ought* is just the same as *is* and *is*, is the realm of empirical fact. . . . This confusion of what *is* with what *should be* is the main problem with Skinner's approach to values.

His rather diffident attempt to get out of this box is evolutionary: 'The survival of a culture then emerges as a new value to be taken into account *in addition to personal and social goods* [emphasis added].' In other words, 'survival'—of the culture or the species—is offered as a superordinate value from which all others can be deduced. This is the source of values Skinner explicitly defended in his utopian writings." John Staddon, *The New Behaviorism: Foundations of Behavioral Science* (New York: Routledge, 2021), 177–178.

30. Skinner, *Beyond Freedom and Dignity*, 128–129, 134–135.

31. Skinner, *Beyond Freedom and Dignity*, 141.

32. See B. F. Skinner, *Walden Two* (New York: The Macmillan Company, 1948). As has been observed, this novel reflects Skinner's struggle to reconcile two aspects of his work: "The struggle was whether to be solely a scientist or also a futuristic prophet with the mission of restructuring society. This struggle is played out in the novel by two main characters, Burris, an academic scholar, and Frazier, the founder of the Utopian community of Walden Two. At the conclusion of the novel, Burris joins Frazier in Walden Two, thus reflecting Skinner's resolution of his personal struggle with the decision to be a futuristic prophet." Robert W. Proctor and Daniel J. Weeks, *The Goal of B. F. Skinner and Behavior Analysis* (New York: Springer, 1990), 102.

33. Skinner, *Beyond Freedom and Dignity*, 148.

34. Skinner, *Beyond Freedom and Dignity*, 176–177.

35. For more detail on this aspect of Darwin's theory, see Davide Tarizzo, *Life: A Modern Invention* (Minneapolis: University of Minnesota Press, 2017), 151–155.

36. Skinner, *Beyond Freedom and Dignity*, 9–10.

37. "The government of Walden Two has the virtues of democracy, but none of the defects. It's much closer to the theory or intent of democracy than the actual practice in America today. The will of the people is carefully ascertained. We have no election campaigns to falsify issues or obscure them with emotional appeals, but a careful study of the satisfaction of the membership is made. Every member has a direct channel through which he may protest to the Managers or even the Planners. And these protests are taken as seriously as the pilot of an airplane takes a sputtering engine. We don't need laws and a police force to compel a pilot to pay attention to a defective engine. Nor do we need laws to compel our Dairy Manager to pay attention to an epidemic among his cows. Similarly, our Behavioral and Cultural Managers need not be compelled to consider grievances. A grievance is a wheel to be oiled, or a broken pipeline to be repaired." Skinner, *Walden Two*, 269.

38. Foucault, for instance, mentions only cursorily Skinner's "behavioral techniques" in his seminar on the birth of biopolitics: "You find these methods in their purest, most rigorous, strictest or aberrant forms, as you wish, in Skinner, and precisely they do not consist in analyzing the meaning of different kinds of conduct, but simply in seeing how, through mechanisms of reinforcement, a given play of stimuli entail responses whose systematic nature can be observed and on the basis of which other variables of behavior can be introduced. In fact, all these behavioral techniques show how psychology understood in these terms can enter the definition of economics given by Becker." Michel Foucault, *The Birth of Biopolitics: Lectures at the Collège de France, 1978–1979* (New York: Palgrave Macmillan, 2008), 270.

39. See Richard H. Thaler and Cass R. Sunstein, *Nudge: Improving Decisions about Health, Wealth, and Happiness* (New Haven, CT: Yale University Press, 2008).

40. "Nudge theory is often described as a multidisciplinary approach to the applied science of human behaviour. In essence, nudge theory suggests that positive reinforcement of behaviours, coupled with hints and suggestions, can (subconsciously) influence motivation, collaboration, and decision processes. What is more is that such nudges towards the 'right' behaviour can often be more effective, and less prone to resistance from groups or individuals, than direct instruction or overt enforcement. The key to successful nudging often involves the individual being unaware that their thoughts, decisions and subsequent behaviours are being influenced by an external force." Craig Jackson, "Behavioural Nudge Theory: A Tool to Improve Safety Behaviour," *Management of Health Risks Special Report* 169 (2013): 1.

41. For more detail on positive and negative reinforcements within the framework of China's social credit systems, see, for instance, Genia Kostka, "China's Social Credit System and Public Opinion: Explaining High Levels of Approval," *New Media & Society* 27, no. 7 (2019): 1565–1593. As Kostka remarks, the social credit systems in China seem to be part of a broader plan to automate social management via feedback loops; that is, cycles of shaping, managing, and responding: "Such mechanisms of positive and negative reinforcements are intended to create a citizenry that continually engages in automatic self-monitoring of its behavior in a manner reminiscent of Foucauldian governmentality" (1568). It is not Foucault, however, but rather Skinner that should be mentioned here. As has been pointed out fairly recently, "the psychological mechanisms of punishments and rewards in the Social Credit System have a striking similarity to the Weltbild of Burrhus F. Skinner's 'behaviourism,' the once famous learning-theory paradigm of 'operant' (or 'instrumental') conditioning that dominated American academic psychology between 1930 and 1960." Martin Woesler, Martin Warnke, Matthias Kettner, and Jens Lanfer, "The Chinese Social Credit System: Origins, Political Design, Exoskeletal Morality and Comparison to Western Systems," *European Journal of Chinese Studies* 2 (2019): 23.

42. Noam Chomsky, Lecture Given at the Institute of Cognitive Science, Carleton University, April 8, 2011: www.youtube.com/watch?v=XbjVMqOk3uc&t=0s.

43. Shoshana Zuboff, *The Age of Surveillance Capitalism: The Fight for a Human Future at the New Frontier of Power* (New York: Public Affairs, 2019), 279–280.

44. See, for instance, Audrey Watters, *Teaching Machines: The History of Personalized Learning* (Cambridge, MA: MIT Press, 2021).

45. See Michel Foucault, *Discipline and Punish: The Birth of the Prison* (New York: Random House, 1977).

46. See, for instance, Charles R. Gallistel, *The Organization of Learning* (Cambridge, MA: MIT Press, 1993); Charles R. Gallistel and John Gibbon, *The Symbolic Foundations of Conditioned Behavior* (Mahwah, NJ: Lawrence Erlbaum Associates, 2002).

47. "Contingencies of selection necessarily lie in the past; they are not acting when their effect is observed. To provide a current cause it has therefore been assumed that they are stored (usually as 'information') and later retrieved. Thus, (1) genes and chromosomes are said to 'contain the information' needed by the fertilized egg in order to grow into a mature organism. But a cell does not consult a store of information in order to learn how to change; it changes because of features which are the product of a history of variation and selection, a product which is not well represented by the metaphor of storage. (2) People are said to store information about contingencies of reinforcement and retrieve it for use on later occasions. But they do not consult copies of earlier contingencies to discover how to behave; they behave in given ways because they have been changed by those contingencies. The contingencies can perhaps be inferred from the changes they have worked, but they are no longer in existence." Skinner, "Selection by Consequences," 58–59.

48. "*I accuse* cognitive scientists of misusing the metaphor of storage. The brain is not an encyclopedia, library, or museum. People are changed by their experiences; they do not store copies of them as representations or rules. *I accuse* cognitive scientists of speculating about internal processes which they have no appropriate means of observing. Cognitive science is premature neurology." B. F. Skinner, "Cognitive Science and Behaviorism," in *Upon Further Reflection* (Englewood Cliffs, NJ: Prentice-Hall, 1987), 111.

49. Skinner, *Beyond Freedom and Dignity*, 196.

50. Skinner, *Beyond Freedom and Dignity*, 194.

51. As a biologist noted some forty years ago, "A glance through natural and social science journals of the last decade makes it obvious that optimization theory is being applied to a variety of problems. . . . Hence optimization does not belong exclusively to biologists, but to any discipline where the subject has some effect on fitness." D. J. Howell, "Optimization of Behavior: Introduction and Overview," *American Zoologist* 23, no. 2 (1983): 257.

52. According to Milton Friedman, "Gary Becker is the greatest social scientist who has lived and worked in the last half century." Quoted in James J. Heckman, *Private Notes on Gary Becker*, IZA Discussion Paper no. 8200 (May 2014): 2.

53. As Becker said during a conversation with François Ewald and Bernard Harcourt, "I don't want to associate with Skinner." Bernard E. Harcourt, Gary S. Becker, and François Ewald, "Becker and Foucault on Crime and Punishment: A Conversation with Gary Becker, François Ewald, and Bernard Harcourt: The Second Session," *Public Law and Legal Theory Working Paper* 440 (2013): 13.

54. Gary S. Becker, *The Economic Approach to Human Behavior* (Chicago: University of Chicago Press, 1976), 294.

55. Becker, *The Economic Approach to Human Behavior*, 7, 14.

56. The concept of "a life" has also been discussed by Barbara Duden, *Die Gene in Kopf—der Fötus im Bauch: Historisches zum Frauenkörper* (Hannover, Germany: Offizin, 2002), 215–252.

57. Paul Rabinow, "French Enlightment: Truth and Life," *Economy and Society* 27, no. 2–3 (1998): 199.

58. For a critique of the idea of individuals qua optimizing agents (both a blessing and a curse for modern economic theory), see Richard H. Thaler, *Misbehaving: The Making of Behavioral Economics* (New York: W. W. Norton, 2016).

59. See B. F. Skinner, "The Evolution of Behavior," in *Upon Further Reflection* (Englewood Cliffs, NJ: Prentice-Hall, 1987), 65–74. Although the concept of modeling can be found in Skinner's works, the name of Albert Bandura is usually associated with it. It is worth noting that Bandura's theory of social learning, highly influential these days, builds on Skinner's research. In the last pages of his major work, for example, Bandura clearly states: "When only a single form of utopian social living is presented as founded on behavioral principles, as in *Walden Two* (Skinner, 1948), the general techniques for developing better social systems get confounded with the particular brand of lifestyle that is promulgated. As a result, procedures for achieving human ideals are repudiated because the advocated mode of life may be uninviting. Principles can be separated from social practices by providing alternative types of social living founded on the same behavioral principles." Albert Bandura, *Social Learning Theory* (Englewood Cliffs, NJ: Prentice-Hall,

1977), 211. In his late writings, Bandura also rephrases Skinner's attack on the "autonomous man" with renewed vigor: "People do not operate as autonomous moral agents. . . . Free will is an enigmatic, autonomous force that is self-negating in function. . . . There is no such thing as autonomous agency that is impervious to environmental influences. Autonomous agency is an illusion." Albert Bandura, *Moral Disengagement: How People Do Harm and Live with Themselves* (New York: Worth Publishers, 2016), 10, 16, 24.

60. No specific reference can be given in this case, for the word "individual" appears in all of Skinner's works.

61. Richard Schacht, *Nietzsche* (New York: Routledge, 1983), 359.

62. Skinner, *Science and Human Behavior*, 9.

63. Davide Tarizzo, *Political Grammars: The Unconscious Foundations of Modern Democracy* (Stanford, CA: Stanford University Press, 2021), 163–199.

64. Cayley, *Ivan Illich in Conversation*, 262.

65. For more details on Julian Huxley and transhumanism, see, for instance, Andrew Pilsch, *Transhumanism: Evolutionary Futurism and the Human Technology of Utopia* (Minneapolis: University of Minnesota Press, 2017).

66. This definition of Life in terms of digital information can be found in Richard Dawkins, *Rivers out of Eden: A Darwinian View of Life* (New York: Basic Books, 1996), 19.

67. Hans Moravec, *Mind Children: The Future of Robot and Human Intelligence* (Cambridge, MA: Harvard University Press, 1988), 2.

68. Ivan Illich, "Health as One's Own Responsibility: No Thank You!" Speech given in Hannover, Germany, on September 14, 1990.

69. Ernst Cassirer, *Language and Myth* (New York: Dover Publications, 1953), 91–92, 94.

70. See Tarizzo, *Political Grammars*, 163–199.

71. Cicero, *The Nature of Gods,* trans. by P. G. Walsh (Oxford, UK: Clarendon Press, 1997), 124.

72. The phrase "reversion to mythology" is taken from Max Horkheimer and Theodor W. Adorno, *Dialectic of Enlightenment* (London: Verso, 1979), 27.

73. Sigmund Freud, *Introductory Lectures on Psycho-Analysis*, in *The Standard Edition of the Complete Psychological Works of Sigmund Freud. Volume XVI (1916–1917)*, ed. James Strachey (London: Hogarth Press, 1963), 414.

74. Freud, *Introductory Lectures on Psycho-Analysis*, 414.

75. In *Walden Two*, in which we find Skinner's description of a perfectly working biopolitical society, we also find a striking illustration of the idea that human beings can be reduced to fully automated living machines. In such a utopian society: "we never encounter any drunkard, any delinquent, any depressive person. We are shown no jail, no psychiatric hospital, not because organised guided tours would be strictly confined to places of interest officially defined as such for visitors, but simply because there are none. Where does that ideal harmony come from?" Marc N. Richelle, *B. F. Skinner: A Reappraisal* (New York: Routledge, 2016), 194.

76. Johan Huizinga, *In the Shadow of Tomorrow* (New York: W. W. Norton, 1964), 223.

Conclusion

1. Barbara Duden, *Disembodying Women: Perspectives on Pregnancy and the Unborn* (Cambridge, MA: Harvard University Press, 1993), 102–103.

2. Michel Foucault, *The Order of Things: An Archaeology of the Human Sciences* (London: Routledge, 1989), 139.

3. Duden, *Disembodying Women*, 104.

4. Duden, *Disembodying Women*, 104.

5. Duden, *Disembodying Women*, 104.

6. Duden, *Disembodying Women*, 73.

7. Duden, *Disembodying Women*, 70.

8. Duden, *Disembodying Women*, 96.

9. Duden, *Disembodying Women*, 92.

10. Duden, *Disembodying Women*, 61.

11. Duden, *Disembodying Women*, 77–78.

12. Duden, *Disembodying Women*, 70.

13. Duden, *Disembodying Women*, 49.

14. Duden, *Disembodying Women*, 59.

15. Duden, *Disembodying Women*, 77–78.

16. Duden, *Disembodying Women*, 60–61.

17. When it comes to defining the sacrum, Duden makes reference to Mircea Eliade: "According to Mircea Eliade, the *sacrum* or *to hieron* is that object in which the transcendent appears. It is a technical term for those things to which the manifestation is tied. The quality and form in which the sacred can be experienced know no limits, but a *sacrum*, a material object, will always be found at the center of culture. In, through, or around this object, the culture's hierophany takes place. Here, the rootedness of reality in the beyond can be experienced. . . . Several anthropologists have accepted Eliade's ideas, some postulating a *sacrum* as a necessary condition for the existence of a culture. They argue that a society rests on the existence of normative assumptions that are neither right nor wrong, true or untrue. These assumptions appear categorically in the *sacrum* without losing their transcendence." Duden, *Disembodying Women*, 108.

18. Duden, *Disembodying Women*, 74.

19. Duden places great emphasis on the role played by the media, as though the latter were the only responsible for such a confusion: "If the fetus is the *sacrum* of our time, it is a *sacrum* of a new kind. In our world, where we increasingly live not among things we see but among appearances we are shown, the modern *sacrum* also has the character of a media event." Duden, *Disembodying Women*, 109.

20. Duden, *Disembodying Women*, 110.

21. When I started my research on transableism, twenty years ago, I found this chronicle written by Robert Vickers on a website dedicated to discussions among and about transabled people (http://biid-info.org/A_date_to_remember). The site was closed shortly afterward. Vickers then recorded a program for ABC Radio National's *Ockham's Razor*, in which he talked about his experience. See https://mindhacks.com/2009/07/13/the-neuroscience-of-an-unwanted-limb.

22. See Michael B. First and Carl E. Fisher, "Body Integrity Identity Disorder: The Persistent Desire to Acquire a Physical Disability," *Psychopathology* 45, no. 1 (2012): 3–14.

23. Rianne M. Bloom, Nienke C. Vulink, Sija J. van der Wal, et al., "Body Integrity Identity Disorder Crosses Culture: Case Reports in the Japanese and Chinese Literature," *Neuropsychiatric Disease and Treatment* 12 (2016): 1419–1423.

24. Michael B. First, "Desire for Amputation of a Limb: Paraphilia, Psychosis, or a New Type of Identity Disorder," *Psychological Medicine* 34 (2004): 7.

25. First, "Desire for Amputation of a Limb," 8.

26. First, "Desire for Amputation of a Limb," 9.

27. Claire Dyer, "Surgeon Amputated Healthy Legs," *British Medical Journal* 320, no. 7231 (2000): 332.

28. See, for instance, Annemarie Bridy, "Confounding Extremities: Surgery at the Medico-Ethical Limits of Self-Modification," *Journal of Law, Medicine & Ethics* 32 (2004): 148–158; Sabine Müller, "Body Integrity Identity Disorder: Is the Amputation of Healthy Limbs Ethically Justified?" *American Journal of Bioethics* 9, no. 1 (2009): 36–43; Annemarie Bridy, "Response to Müller," *American Journal of Bioethics* 9, no. 9 (2009): W8; Christopher James Ryan, "Out on a Limb: The Ethical Management of Body Integrity Identity Disorder," *Neuroethics* 2 (2009): 21–33; Daniel Patrone, "Disfigured Anatomies and Imperfect Analogies: Body Integrity Identity Disorder and the Supposed Right to Self-Demanded Amputation of Healthy Body Parts," *Journal of Medical Ethics* 35, no. 9 (2009): 541–545; Christopher James Ryan, Tarra Shaw, and Anthony W. F. Harris, "Response to Patrone: Body Integrity Identity Disorder," *Journal of Medical Ethics* 36, no. 3 (2010): 189–190.

29. "*Gender dysphoria* refers to the distress that may accompany the incongruence between one's experienced or expressed gender and one's assigned gender. Although not all individuals will experience distress as a result of such incongruence, many are distressed if the desired physical interventions by means of hormones and/or surgery are not available. The current term is more descriptive than the previous DSM-IV term *gender identity disorder* and focuses on dysphoria as the clinical problem, not identity per se." American Psychiatric Association, *Diagnostic and Statistical Manual of Mental Disorders, Fifth Edition* (Arlington, VA: American Psychiatric Association, 2013), 451. See also Allan V. Horwitz, *DSM: A History of Psychiatry's Bible* (Baltimore: John Hopkins University Press, 2021), 140–141.

30. Tim Bayne and Neil Levy, "Amputees by Choice: Body Integrity Identity Disorder and the Ethics of Amputation," *Journal of Applied Philosophy* 22, no. 1 (2005): 75–86.

31. Amy White, "Body Integrity Identity Disorder beyond Amputation: Consent and Liberty," *HEC Forum* 26, no. 3 (2014): 225–236.

32. As has been observed lately, "BIID is a rare but intriguing condition that is yet to be fully understood. The conceptual framework for investigating it has evolved over the past 40 years, from a description that privileged the paraphilic aspects of the clinical features to more recent accounts that favour neuropsychological and neuropsychiatric processes that are thought to be manifestations of dysfunction in the right superior parietal lobule. There is undoubted moral discomfort, if not repugnance, at the notion of elective amputation of a healthy limb but this may be partially modified by a realisation that this unusual phenomenon is a reflection of underlying pathophysiology. There is, in our view, no logical difference between the conceptual status of BIID and transsexualism. Hence, given that individuals with transsexualism are offered gender reassignment surgery it seems to us that individuals with BIID ought at least to be considered for treatment, including elective amputation in some cases. It may be that the need for this radical and controversial form of treatment would soon be obviated by novel treatments involving multimodal sensory stimulation. There is no doubt though that BIID is a morally challenging condition and that the use of surgical amputation or transection of the spine as treatment is deeply troubling." Emma Barrow and Femi Oyebode, "Body Integrity Identity Disorder: Clinical Features and Ethical Dimensions," *BJPsych Advances* 25 (2019): 193–194.

33. Among the opponents of surgical amputation, many place a heavy emphasis on the Judeo-Christian-Islamic roots of our civilization: "As opposed to those who argue for the dispensability of the human body, the Judeo-Christian-Islamic tradition affirms the sanctity of the human body. This is reflected in the ritual care attended to the human corpse culminating in a dignified burial. In many traditions, burial is also required of an isolated limb attesting to its inherent sanctity. In Judaism, respect for the body is also related to the theological concept of the dignity of man who was created in the image of God. The human body designed by God confers dignity on man, and a mutilated body is an affront to this God given gift. For this reason, the ancient practice of mutilation associated with mourning or tattooing as a religious practice was forbidden by Judaism. These sentiments also explain why many adherents of these traditions are incensed by the use of body parts for artistic purposes. Hinduism and other Eastern religions also invest the body with a form of holiness and deliberate maiming of the body is sacrilegious in these ancient traditions. For these reasons, other alternative treatments should be offered to patients who have BIID." Alan Jotkowitz and Ari Zivotofsky, "Body Integrity Identity Disorder (BIID) and the Limits of Autonomy," *American Journal of Bioethics-Neuroscience* 9, no. 1 (2009): 56.

34. First and Fisher, "Body Integrity Identity Disorder," 3–14.

35. See, for instance, Anna Sedda, "Body Integrity Identity Disorder: From a Psychological to a Neurological Syndrome," *Neuropsychology Review* 21, no. 4 (2011): 334–336; Anna Sedda and Gabriella Bottini, "Apotemnophilia, Body Integrity Identity Disorder or Xenomelia? Psychiatric and Neurologic Etiologies Face Each Other," *Neuropsychiatric Disease and Treatment* 10 (2014): 1255–1265.

36. Georg Wilhelm Friedrich Hegel, *Elements of the Philosophy of Right*, ed. Allen W. Wood (Cambridge: Cambridge University Press, 1991 [orig. 1821]), 37–38.

37. John Stuart Mill, *On Liberty*, ed. David Bromwich and George Kateb (New Haven, CT: Yale University Press, 2003 [orig. 1859]), 81.

38. Joel Feinberg, "Legal Paternalism," *Canadian Journal of Philosophy* 1, no. 1 (1971): 120. See also Joel Feinberg, *Harm to Self: The Moral Limits of the Criminal Law* (Oxford: Oxford University Press, 1986).

39. Feinberg, "Legal Paternalism," 111. For objections to Feinberg's argument, see Richard Arneson, "Joel Feinberg and the Justification of Hard Paternalism," *Legal Theory* 11, no. 3 (2005): 269–284.

40. See, for instance, Robin Mackenzie and Stephen Cox, "Transableism, Disability and Paternalism in Public Health Ethics: Taxonomies, Identity Disorders and Persistent Unexplained Physical Symptoms," *International Journal of Law in Context* 2, no. 4 (2006): 363–375.

41. Stefano Rodotà notes that "in cases such as these, the social answer cannot be limited to passively recording the effects of a pathological condition, thus hastily freeing ourselves of a dramatic problem." Stefano Rodotà, *La vita e le regole. Tra diritto e non diritto* (Milan: Feltrinelli, 2006), 88–89.

42. See Sigmund Freud, *The "Uncanny,"* in *The Standard Edition of the Complete Psychological Works of Sigmund Freud. Volume XVII (1917–1919)*, ed. James Strachey (London: Hogarth Press, 1955), 217–256.

43. Georg Wilhelm Friedrich Hegel, *The Phenomenology of Spirit*, ed. Terry Pinkard (Cambridge: Cambridge University Press, 2018 [orig. 1807]), 255. See also Terry Pinkard, "What Is a Shape of Spirit?" in *Hegel's Phenomenology of Spirit*, ed. Dean Moyar and Michael Quante (Cambridge: Cambridge University Press, 2008), 112–129.

44. Hegel, *The Phenomenology of Spirit*, 255.

45. Hegel, *The Phenomenology of Spirit*, 360.

46. Hegel, *The Phenomenology of Spirit*, 255.

47. Wolfgang Schäuble, "Interview," *Der Tagesspiegel*, April 26, 2020.

48. The scholarly literature on this topic is rapidly growing but still rudimentary. See, for instance, Julian Arato, Kathleen Claussen, and J. Benton Heath, "The Perils of Pandemic Exceptionalism," *American Journal of International Law* 114, no. 4 (2020): 627–636; Gerard Delanty, ed., *Pandemics, Politics, and Society: Critical Perspectives on the Covid-19 Crisis* (Berlin and Boston: De Gruyter, 2021); Dave Cowan and Ann Mumford, eds., *Pandemic Legalities: Legal Responses to Covid-19—Justice and Social Responsibility* (Bristol: Bristol University Press, 2021); Paulo Vila Maior and Isabel Camisão, *The Pandemic Crisis and the European Union: Covid-19 and Crisis Management* (London: Routledge, 2022).

49. For a brief analysis of what science has become during the COVID-19 pandemic, see John P. A. Ioannidis, "How the Pandemic Is Changing the Norms of Science," www.tabletmag.com, September 9, 2021.

50. Hannah Arendt, *The Life of the Mind. One-Volume Edition* (New York: Harcourt, 1978), 88.

51. Arendt, *The Life of the Mind*, 97.

52. In Italian and other languages, the confusion was all the more evident because of the clear lexical distinction that can be made between *contagi* (infection cases) and *casi* (disease cases). At some point, public authorities and the media began to speak of *contagi* in terms of *casi*.

53. Lee Clarke, *Worst Cases: Terror and Catastrophe in the Popular Imagination* (Chicago: University of Chicago Press, 2006), 162.

54. Clarke, *Worst Cases*, xi.

55. See Michael Brendan Dougherty, "Anthony Fauci: I Am the Science," *National Review* (November 29, 2021): www.nationalreview.com/2021/11/anthony-fauci-i-am-the-science.

56. B. F. Skinner, *Beyond Freedom and Dignity* (London: Penguin Books, 1971), 155, 162, 170.

57. Skinner, *Beyond Freedom and Dignity*, 134–135.

58. The naturalistic fallacy has always been a thorny issue for evolutionary biologists, sociobiologists, and philosophers of biology. See, for instance, Michael Ruse and Edward O. Wilson, "Moral Philosophy as Applied Science," *Philosophy* 61, no. 236 (1986): 173–192; Michael Ruse, "Evolution and the Naturalistic Fallacy," in *The Naturalistic Fallacy*, ed. Neil Sinclair (Cambridge: Cambridge University Press, 2019), 96–116. For an overview of current debates on evolutionary ethics and metaethics, see Michael Ruse and Robert J. Richards, eds., *Cambridge Handbook of Evolutionary Ethics* (Cambridge: Cambridge University Press, 2017).

59. Samuel Butler, *Erewhon, or, Over the Range* (London: Trübner and Co., 1872), 85.

60. Skinner, *Beyond Freedom and Dignity*, 76.

61. Reflections on Artificial Life can be found in Samuel Butler, "Darwin among the Machines," *The Press*, June 13, 1863; Butler, *Erewhon*, 190–223. For a discussion of Darwin's theory and its teleological implications, see Samuel Butler, *Evolution, Old and New; or, The Theories of Buffon, Dr. Erasmus Darwin, and Lamarck, as Compared with that of Mr. Charles Darwin* (Salem, MA: S. E. Cassino Publisher, 1879), 345–372.

62. See Kenneth J. Rothman, *Epidemiology: An Introduction* (Oxford: Oxford University Press, 2012), 43–44.

63. Immanuel Kant, *The Metaphysics of Morals*, ed. Mary Gregor (Cambridge: Cambridge University Press, 1991 [orig. 1797]), 63.

64. Hegel, *Elements of the Philosophy of Right*, 21.

65. Norberto Bobbio, *The Age of Rights* (Cambridge, UK: Polity Press, 1996).

66. Hegel, *Elements of the Philosophy of Right*, 26.

67. See Davide Tarizzo, *Political Grammars: The Unconscious Foundations of Modern Democracy* (Stanford, CA: Stanford University Press, 2021), 209–210.

68. See, for instance, Alfred Rosenberg, *Der Mythus des 20. Jahrhunderts: Eine Wertung der seelisch-geistigen Gestaltenkämpfe unserer Zeit* (Munich: Hoheneichen, 1930).

Index of Names

Abramovitch, Henry, 225
Abrams, Peter, 232
Adam, Barbara, 243
Adorno, Theodor W., 257
Agamben, Giorgio, 89, 238, 240
Alexopoulos, George S., 248
Alford, John R., 241
Allen, Harold J., 248
American Psychiatric Association, 124, 259
Anderson, Michael, 132–133
Arato, Julian, 261
Arendt, Hannah, 209–210, 261
Ariès, Philippe, 3–5, 226
Arieti, Silvano, 236
Ariew, André, 229, 231
Aristotle, 6, 75–76, 187, 206, 226
Arneson, Richard, 260
Atsbeha, Berhanemeskel Weldegerima, 250
Ayala, Francisco J., 231

Baird, Scarlett O., 245
Balibar, Étienne, 247
Baltimore, David, 236
Bandura, Albert, 245, 256–257
Barnes, Barry, 237
Barrett, Paul, 228, 230
Barrow, Emma, 260
Bayne, Tim, 259
Beck, Ulrich, 243
Becker, Ernest, 2, 225
Becker, Gary S., 133, 167–170, 250, 254, 256
Beddoes, Thomas, 228
Belay, Wudneh Simegn, 250
Benjamin, Walter, 61–62, 234
Bentham, George, 41, 43, 232
Bentham, Jeremy, 118, 163–164, 169
Berlinski, David, 233
Birhanu, Biruktawit Shewafera, 250
Blank, Robert H., 241
Blashfield, Roger K., 248
Bloom, Rianne M., 259
Blumenbach, Johann Friedrich, 58, 234
Bobbio, Norberto, 221, 262
Boncinelli, Edoardo, 79–80, 238, 239
Bostrom, Nick, 250
Bottini, Gabriella, 260
Boudry, Maarten, 236
Boyd, Robert, 252
Braidotti, Rosi, 240
Bridgman, Percy W., 248
Bridy, Annemarie, 259
Broadbent, Alex, 243
Bröckling, Ulrich, 243
Buffon, Georges-Louis Leclerc de, 31–32, 230
Buka, Stephen L., 248
Burdach, Karl Friedrich, 228
Burgess, Danny R., 248
Burian, Richard M., 233, 234
Burkeman, Oliver, 244
Butler, Samuel, 215–216, 262

Cahalan, Susannah, 247
Camisão, Isabel, 261
Canguilhem, Georges, 16, 113, 228, 246
Carus, Carl Gustav, 23
Cassirer, Ernst, 178–179, 257
Castel, Robert, 243
Cayley, David, 251, 257
Chambers, Robert, 59
Chase, Jonathan M., 227
Chomsky, Noam, 162, 243, 255
Chorpita, Bruce F., 245
Cicero, 181, 257

Clarke, Lee, 261
Claussen, Kathleen, 261
Cleland, Carol, xi, 225
Coleman, William, 228
Coleridge, Samuel, 10, 226
Comfort, Nathaniel, 237
Cook-Deegan, Robert, 250
Cooper, Melinda, 240
Cowan, Dave, 261
Cox, Stephen, 260
Crick, Francis H., 81, 239
Cuvier, Georges, 58, 234

Darwin, Charles, xiv, 2, 6–8, 11–18, 23–37, 39–47, 50–53, 58–61, 66–68, 71, 78–81, 118, 145, 148, 156, 160, 166, 215–216, 226–235, 238–239, 254
Darwin, Erasmus, 24
Davies, Paul Sheldon, 234
Davis, Michelle L., 245
Dawkins, Richard, 54, 56, 60–61, 81, 227, 232, 233, 235, 239, 257
De Caro, Mario, 236
Delanty, Gerard, 261
Deleuze, Gilles, 109–111, 114, 120–121, 246, 247
Demeke, Chilot Abiyu, 250
Depew, David, 228,
Dewey, John, 7, 226
Dobson, Keith S., 244
Dobzhansky, Theodosius, 24, 227, 228–229
Doll, Richard, 97, 99, 243
Donnelly, Kevin, 247
Dougherty, Michael Brendan, 261
Dozois, David J. A., 244
Duden, Barbara, 186–195, 240, 256, 258
Duns Scotus, John, 78
Dupré, John, 235, 237, 239
Duve, Freimut, 195
Dyer, Claire, 259

Ehrenberg, Alain, 250
Einstein, Albert, 47, 241
Eldredge, Niles, 15, 46–47, 227, 232, 238
Eliade, Mircea, 258
Ellis, Albert, 244
Ereshesfky, Marc, 235
Esposito, Roberto, 240
Ewald, François, 243, 256

Farah, Martha J., 250
Farr, William, 96, 242
Farrell, Nisha, 250
Fauci, Anthony, 261
Feest, Uljana, 248
Feinberg, Joel, 205–206, 260
Feyerabend, Paul, 226
First, Michael B., 197–199, 201–203, 259, 260
Fisher, Carl E., 259, 260
Fisher, Dennis G., 250
Fisher, Ronald, 60–62, 234
Fodor, Jerry, 7–8, 11–12, 226, 235
Foucault, Michel, xv, 8–10, 18–19, 27, 88–89, 91, 92, 95–96, 99, 101–104, 109–115, 117, 121–122, 135–137, 163, 178, 187, 189, 221, 226, 228, 229, 241–243, 245–247, 251, 254–256, 258
Franklin, Sarah, 240
Freud, Sigmund, 183, 203, 233, 257, 261
Friedman, Milton, 256
Fromm, Erich, 175, 194

Galilei, Galileo, 47, 232
Gallistel, Charles R., 255
Galton, Francis, 115
Gans, Carl, 234
Gayon, Jean, 24, 228, 230
Geoffroy Saint-Hilaire, Étienne, 58
Gerhart, John, 235
Ghiselin, Michael T., 235
Gibbon, John, 255
Gilman, Stephen E., 248
Gilson, Étienne, 226
Glauberman, Naomi, 237
Goethe, Johann Wolfgang von, 235
González-Ayesta, Cruz, 238
Gould, Stephen Jay, 41, 118, 141, 231–233, 237, 247, 251
Green, Joseph H., 23
Greenberg, Jeff, 225
Grene, Marjorie, 228, 231

Hanov, Michael Christoph, 228
Haraway, Donna, 240
Harcourt, Bernard E., 256
Hardt, Michael, 111, 240, 246
Harris, Anthony W. F., 259
Haslam, Nick, 248
Heath, J. Benton, 261
Heckman, James J., 256
Hegel, Georg Wilhelm Friedrich, 204, 207–209, 220–221, 260–262
Heidegger, Martin, 226
Hempel, Carl Gustav, 125–126, 230–231, 248
Herder, Johann Gottfried, 28, 229
Herrnstein, Richard J., 74, 115, 237, 246
Herschel, John Frederick William, 41–44, 46, 48, 232
Hertz, Robert, 4–5, 226
Hibbing, John R., 241
Hill, A. Bradford, 97, 99, 108, 243
Hines, Samuel M. Jr., 241
Hodge, M. J. S., 232
Horkheimer, Max, 257
Horwitz, Allan V., 244, 249, 250, 259
Howell, D. J., 256
Huizinga, Johan, 183, 257
Hull, David L., 231, 235
Humboldt, Alexander von, 23
Hutchinson, George Evelyn, 54–55, 233
Huxley, Julian S., 83, 176, 239, 257
Huxley, Thomas Henry, 235

Illes, Judy, 250
Illich, Ivan, 139–140, 175–178, 180, 188, 251, 257
Ioannidis, John P. A., 261

Jackson, Craig, 255
Jacoby, Russell, 237
Jaffe, Adi, 250
Jeste, Dilip, 248
Johnson, Curtis N., 234
Jonas, Hans, 226
Jotkowitz, Alan, 260

Kahneman, Daniel, 94, 100, 242
Kamin, Leon J., 237
Kant, Immanuel, 17–18, 76, 82, 204, 209, 219–221, 238, 239, 262
Karim, Tazin, 250
Kauffman, Stuart A., 226
Kebede, Beza Tefera, 250
Keller, Evelyn Fox, 233, 237
Kerényi, Karl, 75–76, 78, 237, 238
Kettner, Matthias, 255
Kirschner, Marc, 235
Kitcher, Philip, 235
Koonin, Eugene V., 233, 237
Kostka, Genia, 255
Krasmann, Susanne, 243

Lakatos, Imre, 7, 226
Lamarck, Jean-Baptiste, 187, 228
Lanfer, Jens, 255
Langton, Christopher G., 149, 252
Larson, Edward J., 235
Laudan, Larry, 236
Lazzarato, Maurizio, 246
Legendre, Pierre, 114, 246
Lehman, Anthony F., 248
Leibniz, Gottfried Wilhelm von, 49, 232
Leibold, Mathew A., 227
Leigh, Egbert Giles Jr., 234, 235
Lekevicius, Edmundas, 233
Lemke, Thomas, 88–91, 240, 243
Lenoir, Timothy, 234
Lentzos, Filippa, 246
Lenzenweger, Marc F., 249
Lévi-Strauss, Claude, 228, 236
Levy, Neil, 259
Lewens, Tim, 229
Lewontin, Richard C., 141, 237, 251
Lilienfeld, David E., 242
Lincoln, Abraham, 241
Lockett, McKenzie, 225
Lucas, Robert E. Jr., 241
Lukács, György, 226
Lumsden, Charles J., 252

Macarthur, David, 236
Mackenzie, Robin, 260
Malow, Robert, 250
Malthus, Thomas Robert, 32, 44, 230

Margulis, Lynn, 225
Matthen, Mohan, 229
Matthew, Patrick, 58
Mayes, Rick, 249
Maynard Smith, John, xii, xiii, 140, 225, 228, 234, 239, 251
Mayr, Ernst, 40, 44, 45, 104, 226–227, 229–232, 236, 238, 252
Meehl, Paul E., 129–130, 249
Menninger, Karl, 249
Meyer, Arthur, 228
Meyer-Emerick, Nancy, 241
Mill, John Stuart, 205, 260
Miller, Jacques-Alain, 243
Milner, Jean-Claude, 114, 243, 246
Miquel, Paul-Antoine, xi, 225
Mizutani, Satoshi, 236
Monod, Jacques, 14, 16, 40, 45, 227–228, 230–232
Morabia, Alfredo, 242
Morange, Michel, 237
Moravec, Hans, 257
Müller, Johannes, 31, 239
Müller, Sabine, 259
Mumford, Ann, 261
Murray, Charles, 74, 115, 237, 246
Muth, John A., 241

Nagel, Ernest, 231
Neander, Karen, 77, 238
Negri, Antonio, 89, 111, 240, 246
Newell, Allen, 148, 252
Newton, Isaac, 11, 25, 43, 47, 232
Nietzsche, Friedrich, 174, 210

Orzack, Steven Hecht, 232, 234
Ospovat, Dov, 235
Owen, Richard, 23
Oyebode, Femi, 260

Palgy, Phyllis, 225
Parascandola, Mark, 242
Patrone, Daniel, 259
Pavlov, Ivan, 162, 244, 245
Pearson, Karl, 35, 37, 46, 108, 115, 230, 245, 246
Pence, Charles H., 232
Pennock, Robert T., 240
Peterson, Steven A., 241
Piattelli-Palmarini, Massimo, 7–8, 11–12, 226, 235
Pico della Mirandola, Giovanni, 28, 229
Pigliucci, Massimo, 236
Pilsch, Andrew, 257
Pittendrigh, Colin, 40, 231
Plomin, Robert, 237
Ploucquet, Wilhelm Gottfried, 189
Poizat, Jean-Claude, 246
Popa, Radu, 225
Popper, Karl, 7, 226, 236
Pörksen, Uwe, 193
Proctor, Robert N., 242
Proctor, Robert W., 254
Putnam, Hilary, 236, 247
Pyszczynski, Tom, 225

Quetelet, Adolphe, 115, 120, 247

Rabinow, Paul, 240, 256
Raoult, Didier, 225, 233
Readings, Bill, 132, 250
Reynolds, Grace L., 250
Ribes-Iñesta, Emilio, 248
Richards, Robert J., 262
Richelle, Marc N., 257
Richerson, Peter J., 252
Robbins, Lionel, 116, 247
Rodotà, Stefano, 261
Rose, Nikolas, 109, 240, 246
Rose, Steven, 237
Rosenberg, Alexander, 252–253
Rosenberg, Alfred, 262
Rosenberg, Rhonda, 250
Rosenhan, David L., 122–125, 127, 247, 248
Rothman, Kenneth J., 242, 262
Ruse, Michael, 228, 230–232, 235, 236, 261, 262
Ryan, Christopher James, 259

Sandberg, Anders, 250
Saracci, Rodolfo, 242
Sargent, Thomas J., 242
Schacht, Richard, 257

Schäuble, Wolfgang, 208, 261
Schelling, Friedrich, 23, 28
Schwarz, Alan, 250
Scitovsky, Tibor, 241
Sedda, Anna, 260
Sellars, Wilfrid, 236
Shaw, Tarra, 259
Simon, Herbert A., 148, 241
Simpson, George Gaylord, 40
Skinner, B. F., 126, 129, 144–174, 183, 214, 215, 219, 221, 240, 248, 251–257, 261, 262
Smith, Adam, 92
Smith, Robert, 199
Smits, Jasper A. J., 245
Snow, John, 96–97
Sober, Elliott, 229, 230, 232, 234–236
Solomon, Sheldon, 225
Somit, Albert, 240, 241
Sonnenschein, Carlos, 237
Soto, Ana M., 237
Spitzer, Robert L., 125–126, 247–249
Staddon, John, 242, 253–254
Stevens, Stanley S., 126, 248
Strickland, Lloyd, 232
Sunder Rajan, Kaushik, 240
Sunstein, Cass R., 161, 254

Tarizzo, Davide, 225, 226, 238, 245, 254, 257, 262
Taylor, Charles, 253
Temin, Howard M., 236
Thaler, Richard H., 161, 241, 254, 256
Traverso, Enzo, 226
Treviranus, Gottfried Reinhold, 6, 228
Tversky, Amos, 94, 100, 242

Valeri, Valerio, 236
van der Wal, Sija J., 259
van Fraassen, Bas, 236
van Loon, Joost, 243
Varmus, Harold, 237
Vickers, Robert, 195, 258
Vila Maior, Paulo, 261
Volungis, Adam M., 244
von Baer, Karl, 59
von Mises, Ludwig, 92, 241
von Neumann, John, xii, 141–145, 148, 225, 251
von Wright, Georg Henrik, 231, 238
Vulink, Nienke C., 259

Wakefield, Jerome C., 244
Walker, Helen M., 246
Walsh, Denis M., 229
Warnke, Martin, 255
Watson, James, 81
Watson, John B., 162, 245
Watters, Audrey, 255
Weed, D. L., 242
Weeks, Daniel J., 254
Weikart, Richard, 226
Weismann, August, 26
White, Amy, 259
Wiegele, Thomas C., 240
Williams, George C., 60, 234
Wilson, Edward O., 146–149, 154, 235, 252, 253, 261
Witcraft, Sara M., 245
Woesler, Martin, 255
World Health Organization (WHO), 133–135, 250, 251
Wright, Sewall, 60, 234

Yimenu, Dawit Kumilachew, 250

Zimmer, Carl, 225
Zivotofsky, Ari, 260
Zubin, Joseph, 125–126
Zuboff, Shoshana, 162, 255